ELEMENTS OF TOPOGRAPHIC DRAWING

ELEMENTS
OF
TOPOGRAPHIC DRAWING

BY

ROSCOE C. SLOANE, C. E.
Professor of Highway Engineering, Ohio State University

AND

JOHN M. MONTZ, B. S. IN C. E.
Associate Professor of Civil Engineering, Ohio State University

SECOND EDITION
FIFTH IMPRESSION

McGRAW-HILL BOOK COMPANY, INC.
NEW YORK AND LONDON
1943

ELEMENTS OF TOPOGRAPHIC DRAWING

COPYRIGHT, 1930, 1943, BY THE
McGRAW-HILL BOOK COMPANY, INC.

PRINTED IN THE UNITED STATES OF AMERICA

*All rights reserved. This book, or
parts thereof, may not be reproduced
in any form without permission of
the publishers.*

THE MAPLE PRESS COMPANY, YORK, PA.

PREFACE TO THE SECOND EDITION

The great advances made in military equipment and practice, air transportation, and railway equipment and service have resulted in many changes and additions in the standard symbols used by these organizations. This edition has two primary objects: first, to revise all symbols to conform to current practice; second, to include a discussion of the use of the polar planimeter and the elements of the common forms of map projection. Other minor changes dictated either by our own teaching experience or by requests of professors teaching the same subject have been added to the subject matter.

The authors wish to acknowledge their indebtedness to the American Railway Engineering Association, the U. S. Coast and Geodetic Survey, the U. S. Geological Survey, the United States Forest Service, the Office of the Chief of Engineers, U. S. War Department, and the Cleveland Regional Underground Survey for their cooperation in furnishing up-to-date material as related to standard symbols.

The authors also wish to make the following personal acknowledgments: to Prof. Raymond E. Davis of the University of California for the paragraph on the Theory of the Polar Planimeter; to Prof. E. F. Coddington and Assoc. Prof. O. J. Marshall of The Ohio State University for discussion and review of the chapter on map projection. Full acknowledgment is also due the Stanford Press of Stanford University, Calif., for the geological symbols as illustrated by John L. Ridgway.

R. C. S.
J. M. M.

COLUMBUS, OHIO.
November, 1943.

PREFACE TO THE FIRST EDITION

The writing of this text was first suggested to the authors by the difficulties encountered in presenting the subject matter to the student and by the diversity of sources from which information may be drawn.

The body of the book has been compiled from mimeographed lectures, illustrations, problems, and practice plates accumulated by the authors in teaching the subject for several years without the aid of a suitable text.

This text is designed as a basis for a course of instruction and as a reference book for the topographic draftsman. To combine under one cover the information necessary for an elementary course in topographic drawing and to include a revised and more elaborate use of symbolic drafting as it has recently been developed by the allied engineering industries have been the chief aims.

The plates and illustrations used have been drawn by the authors themselves, and an attempt has been made to fix the scale so that the book reproductions will show the component strokes clearly.

The authors wish to acknowledge their indebtedness to The Board of Surveys and Maps, Washington, D. C., to the Board of Engineers, War Department, to the American Railway Engineering Association, and to the R. H. Randal Engineering Company, of Toledo, Ohio, for their assistance and cooperation in preparing the text. Acknowledgment is also made to Prof. Thomas E. French, Department of Engineering Drawing, Ohio State University, for many helpful hints and suggestions included in the chapter on Suggestions for Office Practice.

R. C. S.
J. M. M.

COLUMBUS, OHIO.
January, 1930.

CONTENTS

	PAGE
PREFACE TO THE SECOND EDITION	v
PREFACE TO THE FIRST EDITION	vii
LIST OF ILLUSTRATIONS	xi
AUTHORIZED ABBREVIATIONS	xv

CHAPTER I

TOPOGRAPHIC DRAWING. 1

    Classes of Information Shown—Classification—Types of Maps—Scales.

CHAPTER II

CONVENTIONAL SIGNS. 25

    Development of Symbols—Effects of Light and Shade—Composition—Choice and Use of Pens—Rules for Making Symbols—Symbols and Signs Indicating Vegetation, Culture, Soils, Relief and Hydrography—Topographic Sketching in Oblique—Miscellaneous Symbols—Geological Symbols and Sections.

CHAPTER III

TOPOGRAPHIC MAPPING. 90

    Plotting of Points—Plotting of Traverses by Polar Coordinates, Chords and Sines of Half Angles, Tangents, Sines, Tangents and Parallel Ruler, Rectangular Coordinates, Latitudes and Departures—Hints and Precautions on Plotting—Blocking Out the Map—Plotting of Details—Use of Protractors—Checking—Order of Penciling Symbols—Inking Topographic Maps—Map Lettering—Types of Letters Used—Size of Letters—Titles—North Points—Border Lines.

CHAPTER IV

TOPOGRAPHIC DRAWING IN COLORS. 125

    Materials and Instruments—General Rules—Stretching the Paper—Preparation of Tints—Color Values—Laying of Flat and Graded Tints—Blending of Tints—Stippling—Dragging—Conventional Signs and Tints—Shading of Slopes—Durability of Colors—Penciling Details—Lettering—Use of Colored Pencils—Colors Used on Geologic Maps.

CHAPTER V

Contours and Contour Sketching 141

Introduction—Contours—Limits of Topographic Expression—Contour Sketching—Glaciation—Erosion Cycles—Alluvial Fans—Wind Erosion—Mountain Formations—Use of Contour Maps—Special Topographical Problems—Profile from Plan—Site Plans—Determination of Drainage Areas—Contours from Controlling Points—Contours from Profiles—Side Elevation from Contour Map—Route Location—Visibility Problems—Adjusting Highway Surfaces from Contours—Glossary of Topographic Forms.

CHAPTER VI

Copying—Duplication—Reproduction 179

Notes on Copying—Use of Glass-top Table—Pantographs—Proportional Squares—Proportional Dividers—Perspective Projection—Tracing—Duplication—Process Papers, *i.e.*, Sensitized Tracing Cloth—Photostat—Ozalid Dry-print Process Reproductions—Zinc Etchings—City or Cadastral Map Reproduction—U. S. Geological Survey Maps—Drawing for Reproduction—Lettering—Line Drawings—General Rules.

CHAPTER VII

Map Projections 195

Introduction—Relation of Type of Projection to Use of Map—Types of Projection, Stereographic, Orthographic, Gnomonic, Cylindrical, Conic, Mercators, Transverse Mercator, Lambert Conformal, Polyconic—Military Grid System—Types of Projection Used by Various States of United States.

CHAPTER VIII

Suggestions for Office Practice 208

Instruments Not in Common Use—Straight Edges—Parallel Ruler—Proportional Dividers—Special Pens as Contour, Railroad, Rivet, Border, Crow-quill, Payzant, Etc.—Beam Compass—Care of Instruments—Use of Curves—Tinting Drawings—Mounting Display Maps—Making Corrections on Maps and Prints—Aids in Lettering—Stock Titles—Care of Field Notes—Universal Drafting Machine—Polar Planimeter, Theory, Use, Accuracy—Other Types of Planimeters—Drawing Papers, Desirable Qualities, Types, Sizes—Pens.

CHAPTER IX

Suggested Problems 234

Problems in Contour Sketching, Drainage Lines, Earthwork, Stream Crossings and Route Location, Federal Government Organizations Using or Producing Maps.

References 243

Index 247

LIST OF ILLUSTRATIONS

PLATES

PLATE	PAGE
1. Sequence of Strokes	28
2. Vegetation Symbols Magnified	32
3. Relief Symbols	44
4. Special Symbols	45
5. Hydrographic Symbols	46
6. Development of Hill Shading	47
7. Vegetation Symbols	48
8. Vegetation Symbols	49
9. Relief Symbols	50
10. Rock Drawing	51
11. Boundary, Fence and Miscellaneous Symbols	52
12. Works and Structures Symbols	53
13. Works and Structures Symbols	54
14. Hydrographic Symbols	55
15. Hydrographic Symbols	56
16. Nautical Chart Symbols	57
17. Miscellaneous Forest Service Symbols	58
18. Underground Survey Symbols	59
19. Topographic Symbols	60
20. A.R.E.A. Symbols—Pipeline and Wire Line	61
21. A R.E.A. Symbols—Miscellaneous	62
22. A.R.E.A. Symbols—Fences, Highways, Mining	63
23. A.R.E.A. Symbols—Railways, Boundary Lines, Fences	64
24. A.R.E.A. Symbols—Track Fixtures and Accessories	65
25. A.R.E.A. Symbols—Bridges, Buildings and Structures	66
26. A.R.E.A. Symbols—Signboards and Miscellaneous	67
27. A.R.E.A. Symbols—Standard Sections	68
28. Military Symbols	69
29. Military Symbols	70
30. Military Symbols	71
31. Symbols for Military Field Sketching	72
32. Military Symbols	73
33. Military Symbols	74
34. Military Symbols	75
35. Military Symbols	76
36. Gas and Oil Symbols	77
37. Golf Course Symbols	78
38. Air Navigation Symbols	79
39. Highway and Waterway Symbols	80
40. Geological Structure Symbols	81
41. Sections Showing Geological Formations	84
42. Sections Showing Geological Formations	85
43. Suggested Topographic Exercises	86
44. Pen and Ink Topography	87

xii

LIST OF ILLUSTRATIONS

Plate	Page
45. Suggested Exercise in Map Toning.	88
46. Suggested Topographic Exercises.	89
47. Roman Alphabets	121
48. Gothic Alphabets.	122
49. Italicized Roman and Gothic Alphabets.	123

Plate	Page
50. Single Line, Stump, and Marginal Alphabets	124
51. Symbols for Color Drawing	facing 139
52. Symbols for Color Drawing	facing 139
53. Water Color Map	facing 139

Figures

Figure	Page
1. Cadastral or City Map.	3
2. Building Site Map.	5
3. Portion of Hydrographic Map.	7
4. Map Illustrating Golf Topography.	9
5. Real-estate Display Map.	10
6. Real-estate Display Map.	12
7. Portion of Landscape Architect's Map.	13
8. Portion of Typical Contour Map of U.S. Geological Survey *facing*	14
9. Map Illustrating Airport Studies	15
10. Map Showing Typical Underground Survey	17
11. Conversion Scale	21
12. Types of Bar Scales	22
12a. Proportional Scale (Diagonal Scale).	23
13. Water Lining.	37
14. Water Lining.	37
15. Relief Shown by Contours	38
16. Relief Shown by Hachures	40
17. Use of Contour Pen	41
18. Shading Scale.	43
19. Traverse Plotting by Polar Protractor.	92
20. Traverse Plotting by Chords	94

Figure	Page
21. Traverse Plotting by Tangents	95
22. Traverse Plotting by Sines	97
23. Traverse Plotting by Tangents and Parallel Ruler . .	99
24. Traverse Plotting by Coordinates	100
25. Traverse Plotting by Latitudes and Departures . . .	102
26. Paper Protractor	104
27. Ames Protractor	104
28. Paper Protractor	105
29. Rectangular Protractor.	106
30. Examples of Letter Construction	114
31. Slope of Slanting Letters.	114
32. Examples of Letter Construction	114
33. Map Lettering—Political Subdivisions.	115
34. Map Lettering—Political Subdivisions (Towns and Villages).	115
35. Map Lettering—Hydrography	116
36. Map Lettering—Hydrography	116
37. Map Lettering—Hypsography	116
38. Map Lettering—Hypsography	117
38a. Map Lettering—Public Works.	117
39. Title Lettering	118
40. Title Lettering	118

LIST OF ILLUSTRATIONS

Figure	Page
41. Suggested North Points	119
42. Brushes	126
43. Mixing Dishes	126
44. Water Pan	126
45. The Application of Color	130
46. Blending	132
47. Position for Dragging	133
48. Contours	142
49. Glaciation—Mountain	145
49a. Glaciated Valley	146
50. Glaciated Area—Flat Country	147
51. Glaciation—Drumlins	148
52. Eroded Valley—First Stage	149
53. Erosion Cycles	150
54. Stages in the Formation of Alluvial Fans	152
55. Contours and Sand Country Sculptured by Wind	153
56. Volcanism	154
57. Construction of Profile from Plan	156
58. Site Plan	158
59. Drainage Area, Flooded Area, and Dam	159
60. Contours Plotted from Control Points	161
61. Checkerboard Map	162
62. Side Elevation from Contour Map	164
63. Plan and Profile of Highway	166
64. Plan and Profile of Paper Location (Proposed Highway)	167
65. Visibility Problems	170
66. Contour Adjustment of Highway Surfaces	171
67. Topographic Forms	174
68. Topographic Forms	175
69. Topographic Forms	176
70. Topographic Forms	177
71. Topographic Forms	178
72. Glass-top Table	180
73. Pantograph	180
74. Pantograph	180
75. Suspended Pantograph	181
76. Enlargement by Squares	182
77. Proportional Dividers	183
78. Reduction by Perspective Projection	184
79. Stereographic Projection	198
80. Orthographic Projection (Polar)	198
81. Gnomonic Projection (Polar)	199
82. Gnomonic Projection (Equatorial)	199
83. Cylindrical Projection	200
84. Mercators' Projection	201
85. Transverse Mercator Projection	201
86. Conic Projection (Tangent Cone)	202
87. Lambert Conformal Projection	202
88. Developed Polyconic Projection Showing Scale Distortion	203
89. Grid Zones—U.S. Military Grid	204
90. Military Grid—Zone "C"	205
91. Straightedge	208
92. Straightedge	209
93. Scale	209
94. Scale	209
95. Parallel Ruler	210
96. Folding Ruler	210
97. Contour Pen	210

LIST OF ILLUSTRATIONS

Figure	Page
98. Railroad Pen	210
99. Swede or Detail Pen	211
100. Rivet Pen	211
101. Jacknife Pen	211
102. Border Pen	212
103. Double Swivel Pen	212
104. Crow-quill Pen	212
105. Payzant Pen	212
106. Payzant Pen	212
107. Beam Compass	212
108. Irregular Curves	214
109. French Curves	215
110. French Curves	215
111. Flexible Curve	215
112. Assembled Set of Railroad Curves	216
113. Lettering Rule	220
114. Braddock Triangle	220
115. Braddock Triangle	221
116. Lettering Triangle	221
116a. Universal Drafting Machine	222
117. Polar Planimeter	222
118. Polar Planimeter	223
118a. Polar Planimeter	224
118b. Polar Planimeter	224
118c. Polar Planimeter	224
119. Contour Problems	235
120. Contour Crossing Road	236
121. Contour Crossing Railroad	237
122. Mistakes in Contour Interpretation	238
123. Contour Problems	240

AUTHORIZED ABBREVIATIONS
GENERAL ABBREVIATIONS FOR STANDARD MAPS

A.	arroyo	E.	east	L.H.	lighthouse	S.	south
abut.	abutment	Est.	estuary	Long.	longitude	s.	steel
A.	arch	f.	fordable	Mt.	mountain	S.H.	school house
b.	brick	Ft.	fort	Mts.	mountains	S.M.	saw mill
B.S.	blacksmith shop	G.S.	general store	N.	north	Sta.	station
bot.	bottom	gir.	girder	n.f.	not fordable	st.	stone
Br.	branch	G.M.	grist mill	p.	pier	str.	stream
br.	bridge	i.	iron	pk.	plank	T.G.	toll gate
C.	cape	I.	island	P.O.	post office	Tres.	trestle
cem.	cemetery	Jc.	junction	Pt.	point	tr.	truss
con.	concrete	kp.	king-post	q.p.	queen-post	W.T.	water tank
cov.	covered	L.	lake	R.	river	W.W.	waterworks
Cr.	creek	Lat.	latitude	R.H.	roundhouse	W.	west
cul.	culvert	Ldg.	landing	R.R.	railroad	w.	wood
D.S.	drug store	L.S.S.	life saving station				

HYDROGRAPHIC SYMBOLS
ABBREVIATIONS RELATING TO BOTTOMS

bk., black; bu., blue; brk., broken; br., brown; cal., calcareous; Cl., clay; crs., coarse; Co., coral; dk., dark; dec., decayed; fne., fine; fly., flinty; G., gravel; gy., gray; gn., green; gty., gritty; grd., ground; hrd., hard; lrg., large; lt., light; M., mud; Oz., ooze; P., pebbles; rd., red; rky., rocky; rot., rotten; S., sand; Sh., shells; sml., small; sft., soft; spk., speckled; Sp., specks; stk., sticky; stf., stiff; St., stones; str., streaky; vol., volcanic; wh., white; yl., yellow.

Abbreviations Relating to Buoys

B., black; C., can; Ch., checkered; G., green; H. S., horizontal stripes; N., nun; R., red; S., spar; V.S., vertical stripes; W., white; Y., yellow.

Abbreviations Relating to Lights

Alt., alternating; B., blue; E., electric; ev., every; F., fixed; Fl., flash; Fls., flashes; Flg., flashing; G., green; Gp., group; m., miles; min., minutes; Occ., occulting; R., red; Rev., revolving; sec., seconds; Sec., sector; U, unwatched; vis., visible; W., white.

Abbreviations Relating to Lights and Buoys

(F.B.) fog bell; (F.D.) fog diaphone; (F.G.) fog gun; (F.H.) fog horn; (F.S.) fog siren; (F.T.) fog trumpet; (F.W.) fog whistle; (S.B.) submarine fog bell; (R.F.S.) radio fog signal.

General Abbreviations

Bn., beacon; Rk., rock; Wk., wreck; N.R.C., naval radio compass; N.R.S., naval radio station; P.D., position doubtful; P.A., position approximate; E.D., existence doubtful; L.M.P., levee milepost; L.S., levee station; TH., towhead.

SPECIAL MILITARY ABBREVIATIONS

The abbreviations listed below are authorized for use in combat orders, messages, records, and reports in the field. With few exceptions all are written without spacing and periods. Dates, when abbreviated, are written thus: 1 Jan 42. Usually a single word is represented by a single capital letter, or by a capital and a lower case letter.

Adjutant (1st staff section, brigades, and lower units)	S-1
Adjutant General	AG
Adjutant General's Department	AGD
Adjutant's (section)	Adj (Sec)
Advance	Adv
Advance Guard	Adv Gd
Advance Message Center	Adv Msg Cen
Air Base	AB
Air Corps	AC
Aircraft warning service	AWS
Air Defense Command	AD Comd
Air District	A Dist
Airdrome	Adrm
Air Force	AF
Air Force Combat Command	AFCC
Air Intelligence (section) (officer)	A Int (SEC) (O)
Airplane	AP
Airship	Ash
Ambulance (company)	Amb (Co)
Ambulance loading post	ALP
American Expeditionary Forces	AEF
Ammunition (distribution point)	Am (DP)
Animal (animal-drawn)	Anl-d
Antiaircraft	AA
Antiaircraft Artillery	AAA
Antiaircraft Artillery Intelligence Service	AAAIS
Antiaircraft Intelligence Service	AAIS
Antimechanized	AMecz
Antitank	AT

AUTHORIZED ABBREVIATIONS

Armored Force	Armd F
Army Air Forces	AAF
Army Engineer Service	A Engr Serv
Army headquarters	AHQ
Army medical (laboratory)	A Med (Lab)
Army Post Office	APO
Artillery (brigade)	Arty (Brig)
Artillery Information Service	AIS
Assistant chief of staff	AC of S
Assistant chief of staff for military intelligence	G-2
Assistant chief of staff for personnel	G-1
Assistant chief of staff for operations and training	G-3
Assistant chief of staff for supply	G-4
Attack	Atk
Automatic	Auto
Auxiliary	Aux
Aviation	Avn
Axis or axes of signal communication	Ax Sig Com
Baggage	Bag
Bakery (battalion)	Bkry (Bn)
Balloon (group)	Bln (Gp)
Barrage balloon (battalion)	Bar Bln (Bn)
Basic	Bsc
Battalion (combat train)	Bn (C Tn)
Battalion adjutant	S-1
Battalion intelligence officer	S-2
Battalion plans and training officer	S-3
Battalion supply officer	S-4
Battery (commander)	Btry (Comdr)
Battle reconnaissance	B Rcn
Bench mark	BM
Bicycle	Bcl
Bomb service truck	BSTrk
Bombardment (squadron) (group) (wing) (light) (medium) (heavy)	Bomb (Sq)(Gp)(Wg)(L)(M)(H)
Boundary	Bd
Bridge train (heavy) (light)	Br Tn (Hv)(L)
Brigade adjutant	S-1
Brigade headquarters	BHQ
Brigade intelligence office	S-2
Brigade plans and training officer	S-3
Brigade supply officer	S-4
Brigadier general	Brig Gen
Browning automatic rifle	BAR
Caliber	cal
Camouflage (battalion)	Cam (Bn)
Cavalry (brigade) (division)	Cav (Brig) (Div)
Cavalry division headquarters	Cav DHQ
Cemetery	Cem
Center	Cen
Centimeter	cm
Chemical (ammunition train) (battalion) (company) (officer) (regiment) (section)	Cml (Am Tn)(Bn)(Co)(O)(Regt)(Sec)
Chemical agent, nonpersistent	G-NP
Chemical agent, persistent	G-P
Chemical mortar, 4.2-inch	4.2 Cml Mort
Chemical Warfare Service	CWS
Chief of artillery	C of A
Chief of aviation	C of Avn

AUTHORIZED ABBREVIATIONS

Chief Signal Officer	C Sig O	Corps headquarters	CHQ
Chief of Staff	C of S	Crossroads	CR
Circular	Cir	Decontamination	Decon
Class I supplies	Cl I Sup	Dental Corps	DC
Clearing (company)	Clr (Co)	Department	Dept
Coast Artillery Corps	CAC	Depot (battalion)	Dep (Bn)
Coast defense	CD	Deputy chief of staff	DC of S
Coastal frontier	CF	Detachment	Det
Collecting (company)	Coll (Co)	Distant surveillance (section)	DS (Sec)
Column	Clm	Distributing point	DP
Combat (car) (zone)	C (Car)(Z)	Division	Div
Command (car)	Comd (Car)	Division (adjutant) (aviation) (commander) (engineer)	Div (Adj)(Avn)(Comdr)(Engr)
Command post	CP		
Command post exercise	CPX	Division headquarters	DHQ
Commandant	Comdt	Division (headquarters commandant and provost marshal)	Div (Hq Comdt & PM)
Commander in Chief	C in C		
Commanding	Comdg	Division (judge advocate) (munitions officer)	Div (JA)(Mun O)
Commanding general	CG		
Commanding officer	CO	Division ordnance officer	Div OO
Commissary	Comm	Division (quartermaster) (signal officer) (surgeon)	Div (QM)(Sig O)(Surg)
Communication (officer) (squadron) (platoon) (section)	Com (O)(Sq)(Plat)(Sec)		
		Dump (truck)	Dp (Trk)
Communications zone	Com Z	Echelon	Ech
Company (commander) (headquarters)	Co (Comdr)(Hq)	Element	Elm
		Embarkation	Emb
Construction (company) (platoon) (section)	Cons (Co)(Plat)(Sec)	Engineer (battalion) (company) (officer) (train)	Engr (Bn)(Co)(O)(Tn)
Contact party	Con Py	Engineers (combat) (general service)	Engrs (C)(Gen Serv)
Continuous wave	CW	Enlisted men	EM
Convalescent (hospital)	Conv (Hosp)	Entrucking point	EP
Corps of Engineers	CE	Evacuation (hospital)	Evac (Hosp)

AUTHORIZED ABBREVIATIONS

Executive (officer)	Ex (O)
Field (remount depot)	F (Rmt Dep)
Field Artillery (brigade)	FA (Brig)
Field laboratory	F Lab
Field train	F Tn
Fighter	Fi
Fixed	Fxd
Flash ranging	FR
Flight (commander)	Flt (Comdr)
Forward echelon	Fwd Ech
Garage	Gar
Gas officer	Gas O
Gasproof dugout or building	G-PF
General (hospital) (service)	Gen (Hosp)(Serv)
General Headquarters	GHQ
General Headquarters Air Force	GHQ AF
General Service	Gen Serv
General Staff (Corps)	GS (C)
First section	G-1
Second section	G-2
Third section	G-3
Fourth section	G-4
Geological (survey)	Geol (Surv)
Graves registration (battalion) (company)	Gr Reg (Bn)(Co)
Group	Gp
Gun	G
Harbor defense	HD
Headquarters (battery) (company) (troop) (platoon) (section)	Hq (Bty)(Co)(Tr)(Plat)(Sec)
Headquarters commandant	Hq Comdt
Headquarters and headquarters (battery) (company) (troop)	Hq & Hq (Btry)(Co)(Tr)
Headquarters and service (battery) (company) (troop)	Hq & Serv (Btry)(Co)(Tr) or H & S (Btry)(Co)(Tr)
Heavy (ponton)	Hv (Pon)
Heavy weapons (company) (section)	Hv Wpn (Co)(Sec)
High explosive	HE
Highway	Hwy
Holding and reconsignment point	H & RP
Horse-drawn	H-Dr
Horse and mechanized	H & Mecz
Hospital (company) (train)	Hosp (Co)(Tn)
Howitzer	How
Infantry (brigade) (division)	Inf (Brig)(Div)
Information center	IC
Inspector general	IG
Inspector General's Department	IGD
Intelligence (platoon) (section)	Int (Plat)(Sec)
Intelligence officer	S-2 or Int O
Interceptor	I
Judge Advocate General's Department	JAGD
Kitchen	Ki
Labor (detachment) (battalion)	Lbr (Det)(Bn)
Laboratory	Lab
Land mine	LM
Lieutenant (colonel) (general)	Lt (Col)(Gen)
Light (machine gun) (tank) (ponton)	L (MG)(Tk)(Pon)
Lights	Lgts
Line of departure	LD

Line of communication	LC
Livens projector	LP
Machine-gun (company) (troop) (officer)	MG (Co)(Tr)(O)
Main line of resistance	MLR
Main supply road	MSR
Maintenance (section)	Maint (Sec)
Maintenance of equipment	Maint of E
Maintenance party	Maint Py
Maintenance and supply	M & S
Maintenance of way	Maint of W
Major (general)	Maj (Gen)
Material	Mat
Mechanized	Mecz
Medical (regiment) (supply depot)	Med (Regt)(Sup Dep)
Medical Corps	MC
Medical Department	MD
Medium (tank) (ponton)	M (Tk)(Pon)
Message (center)	Msg (Cen)
Message dropping and pick-up ground	Msg DPU
Messenger (section)	Msgr (Sec)
Meteorological (company) (officer) (section)	Met (Co)(O)(Sec)
Military intelligence	MI
Military police (battalion) (company)	MP (Bn)(Co)
Mobilization Regulations	MR
Mobilization Training Program	MTP
Mobile	Mbl
Mortar	Mort
Motor Transport (Service)	MT (S)
Motorcycle	Mtrcl
Motorized	Mtz
Mounted	Mtd
Munitions (officer)	Mun (O)
Navigation (officer)	Nav (O)
Net control station	NCS
Noncommissioned officer	NCO
Observation (battalion) (flash) (group) (squadron)	Obsn (Bn)(Fl)(Gp)(Sq)
Observation post	OP
Observer	Obsr
Officer, order, or orders	O
Operation	Opn
Ordnance	Ord
Ordnance officer	OO
Organization	Orgn
Outpost line	OPL
Outpost line of resistance	OPLR
Pack (troop) (train)	Pk (Tr)(Tn)
Parachute	Prcht
Park	Prk
Party	Py
Personnel (carrier) (officer) (section)	Pers (Car)(O)(Sec)
Photographic (section) (squadron)	Photo (Sec)(Sq)
Pigeon company	Pgn Co
Pioneer	Pion
Pioneer and demolition section	P & D Sec
Plans and training officer	S-3
Platoon (commander) (headquarters)	Plat (comdr)(Hq)
Point	Pt
Ponton	Pon
Postal (section)	Post (Sec)

AUTHORIZED ABBREVIATIONS

Prisoners of war	PW
Private	Pvt
Provisional	Prov
Provost marshal	PM
Provost Marshal General	PMG
Pursuit (group) (squadron)	Pur (Gp)(Sq)
Quartermaster (section)	QM (sec)
Quartermaster Corps	QMC
Radio (company) (section)	Rad (Co)(Sec)
Radio intelligence (company)	Rad Int (Co)
Radio and panel section	R & P Sec
Railhead (detachment)	Rhd (Det)
Railhead officer	RHO
Railroad transportation officer	RTO
Railway (battalion)	Ry (Bn)
Railway traffic officer	R Traf O
Range (officer)	Rg (O)
Ration distributing point	RDP
Rear (echelon) (guard)	Rr (Ech) (Gd)
Reception (center)	Recp (Cen)
Reconnaissance (officer)	Rcn (O)
Reconstruction park	Recons Prk
Regiment	Regt
Regimental	Regtl
Regimental adjutant	S-1
Regimental headquarters	RHQ
Regimental intelligence officer	S-2
Regimental plans and training officer	S-3
Regimental reserve line	RRL
Regimental supply officer	S-4
Regulating (officer) (point) (station)	R (O)(P)(Sta)
Reinforced	Reinf
Remount (depot) (officer) (service)	Rmt (Dep)(O)(Serv)
Repair (section)	Rep (Sec)
Replacement	Repl
Replacement Training Center	RTC
Reproduction	Repro
Reserve	Res
Rifle (company) (platoon) (squad)	R (Co)(Plat)(Sqd)
Road	Rd
Road bend	RB
Road junction	RJ
Road space	RS
Sales commissary (battalion) (company)	Sales Comm (Bn)(Co)
Salvage (battalion) (company)	Salv (Bn)(Co)
Sanitary	Sn
Scheduled	Scd
School	Sch
Scout	Sct
Scout car	Sct C
Scout car, half-track	Sct C Ht
Searchlight	Slt
Section	Sec
Self-propelled mount	SPM
Semiautomatic rifle	SAR
Semimobile	Sem
Service (battalion) (battery) (company) (troop)	Serv (Bn)(Btry)(Co)(Tr)
Signal (battalion) (company) (troop) (depot) (officer) (section)	Sig (Bn)(Co)(Tr)(Dep)(O)(Sec)
Signal Corps	Sig C

Signal operations instructions	SOI
Small arms (ammunition)	SA (Am)
Sound and Flash (battalion)	S & F (Bn)
Sound locator	S-L
Sound ranging	SR
Special troops	Sp Trs
Special weapons (troop) (platoon)	Sp Wpn (Tr)(Plat)
Squadron (headquarters) (headquarters and headquarters detachment)	Sq (Hq)(Hq & Hq Det)
Staff (group) (squadron)	Stf (Gp)(Sq)
Standing operating procedure	SOP
Station	Sta
Straggler line	Strag L
Submachine gun	SMG
Supply (company) (section)	Sup (Co)(Sec)
Supply officer	S-4 or Sup O
Supply point	Sup Pt
Support line	SL
Surgical (hospital)	Surg (Hosp)
Survey	Surv
Switchboard	Sb
Switching central	Sw C
Tank (battalion) (company)	Tk (Bn)(Co)
Tank destroyer	TD
Telegraph (section)	Tg (Sec)
Telephone (section)	Tp (Sec)
Theater headquarters	THQ
Theater of operations	T of Opns
Topographic	Top or Topo
Tractor	Trac
Tractor-drawn	Tr Dr
Traffic	Traf
Train	Tn
Trailer	Tlr
Transport or transportation (platoon)	T (Plat)
Trench mortary (battery)	T Mort (Btry)
Troop (s)	Tr (s)
Truck (battalion) (company) (section) (truck head)	Trk (Bn)(Co)(Sec) (Trk hd)
Truck-drawn	Trk-Dr
Truck head	Trk hd
Veterinary or veterinarian	Vet
Veterinary (company) (evacuation hospital) (service)	Vet (Co)(Evac Hosp) (Serv)
Veterinary Corps	VC
Visual	Vis
Wagon (battalion) (section) (train)	Wag (Bn)(Sec)(Tn)
War Department	WD
Water supply	W Sup
Water tank (battalion) (train)	W Tk (Bn)(Tn)
Weapon carrier	Wpn Carr
Weapons (troop) (platoon)	Wpn (Tr)(Plat)
Weather (group) (officer) (squadron)	Wea (Gp)(O)(Sq)
Wing	Wg
Yard	Yd
Year	Yr
Zone	Z
Zone of the interior	Z of I

ELEMENTS OF TOPOGRAPHIC DRAWING

CHAPTER I

TOPOGRAPHIC DRAWING

Topographic drawing is the art of recording to scale, by means of conventional symbols, the topographic features found on that portion of the earth's surface which the map in question represents. A topographic map differs from a mechanical drawing in that it is a combination of both freehand and mechanical draftsmanship. Its purpose is to convey to the trained eye a mass of information with the greatest possible speed. Conventional symbols of a distinctive form, and frequently of a distinctive color, lend themselves readily to this purpose and can be shown complete in one view. This view in American practice corresponds to the plan view in mechanical drawing. In addition to the use of symbols which constitute the major portion of a topographic map, a considerable amount of lettering is necessary to definitely designate such major features as rivers, lakes, political subdivisions, highways, railroads, etc., by their proper names.

The information shown on a complete topographic map may be divided into four major classifications:

1. Culture, or the works of man.
2. Relief, *i.e.*, relative elevations and depressions.
3. Hydrographic, or water features.
4. Vegetation.

On some topographic maps the information shown is confined to the first three classes, and no attempt is made to indicate the type of vegetation present.

Culture.

Culture includes all features built or constructed by human agency, with the exception of canals, artificial cuts and fills, and similar features which may be more properly classified as either hydrographic or relief features. Culture more properly includes highways, railroads, towns, fences, houses, and all lines indicating political authority or private ownership. The above includes state, county, township, and private-property lines.

Relief.

Topographic features representing variations in elevation of the earth's surface are classified under relief. These variations are indicated by certain conventional symbols or by a combination of symbols and lettering. Valleys, hills, plains, plateaus, and mountains are classified as relief.

Hydrography.

Hydrographic features include all information pertaining to water which may be represented on a topographic map. Oceans, seas, lakes, ponds, rivers, creeks, and other large or small bodies of water are included in this classification.

Vegetation.

This classification includes all such natural features as forests, orchards, crops, meadows, etc., whether occurring naturally or planted by man.

Classes of Topographic Maps Based on Use.

From the standpoint of use, topographic maps may be grouped in the following classes:

1. *Geographic Maps.*—These consist of maps covering a very large area and showing only the most prominent relief, cultural, and hydrographic features, such as large mountain ranges, important cities, state and national boundaries, and large bodies of water. Such maps are generally made on a scale of 1/1,000,000 to 1/62,500, or expressed in another way, on a scale of from 1 in. = about 15 miles to 1 in. = 1 mile.

2. *Cadastral or City Maps.*—This class includes all accurate topographic maps made to serve as a control in the development and growth of a large city or suburban district. They are made on a scale of 1 in. = 1,000 ft. to 1 in. = 200 ft., and constitute a valuable record of the physical facts which control city operation and planning.

The scaling accuracy is about 0.01 in. and the contour interval is taken at 1 to 2 ft., depending upon local conditions.

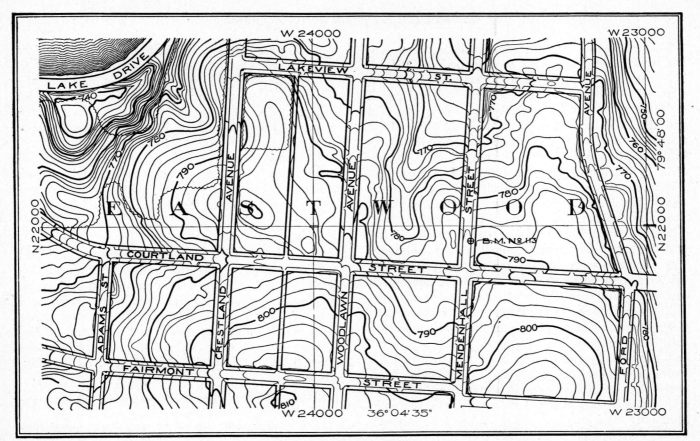

Fig. 1.—Cadastral or city map.

They are based on an accurate triangulation control from which a system of coordinates is calculated. The map is made in sheets corresponding to subdivisions of the coordinate system. The original pencil drawings are the plane table sheets made in the field and should be preserved in field condition, without tracing in ink, as a basis for revisions or additions, as future needs may indicate. Figure 1 shows a portion of a coordinate sheet copied from a city survey. The originals are produced in colors similar in every way to the U. S. Geologic Survey quadrangle sheets except in the items of accuracy, detail, and contour interval. Since the scaling accuracy is small, the drafting must be accurate, neat, and painstaking. The final maps are a basis for planning all future subdivisions, sewer and water extensions, boulevards and streets, and constitute a valuable asset to any city.

3. *General Topographic Maps.*—Under this group may be placed topographic maps made in connection with the development of very large industrial projects and other large engineering developments, such as hydroelectric projects, drainage developments, flood prevention, etc. These maps differ from cadastral maps in that they may not be so accurate, the scale commonly used varying from 1/62,500 or 1 = about 1 mile, to 1/6,000 or 1 in. = 500 ft.

4. *Engineering or Working Maps.*—This class includes the vast number of maps made for use in connection with the preliminary planning and the construction of engineering projects, etc. The scales used vary from 1/2,400 or 1 in. = 200 ft. to 1/600 or 1 in. = 50 ft., although on landscape maps the scale used may be as large as 1/120 or 1 in. = 10 ft.

Working maps vary in amount and nature of information shown according to the final purpose and use of the map. The field is so diversified and examples so numerous that only a few will be shown.

Building Site.

In the preparation of building plans, the architect requires a map of the site showing the information necessary to the proper location of buildings both in plan and elevation. Such maps are drawn to a scale of 1 in. = 10 ft. or 1 in. = 20 ft. These maps show much detail, as shown in Fig. 2, and require considerable lettering. The following specifications cover most of the information required on building-site maps:

1. Show all existing pins or stones marking lot corners as well as those set on last survey.
2. Show all property lines, their lengths and offsets from sidewalks.

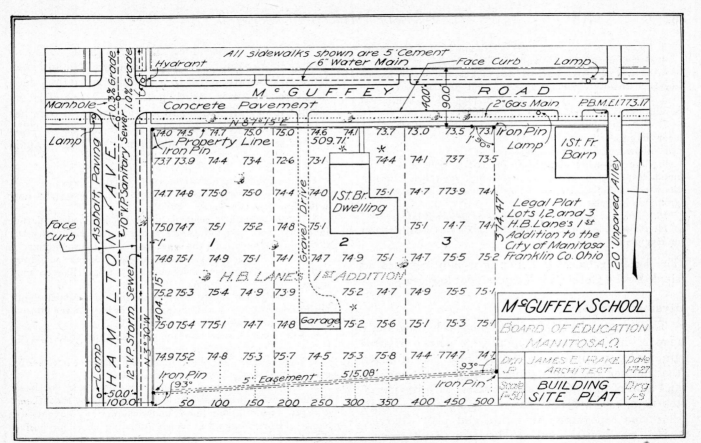

FIG. 2.—Building site map.

3. Show street lines, curb lines, width and kind of pavement.
4. Same information for walks and drives.
5. Show exact location and size of gas and water mains, together with house connections, if already in place.
6. Show location, kind, and size of pipe, of all storm and sanitary sewers, together with all manholes.
7. Show location of all poles, hydrants, conduits, etc. on or near the property.
8. Show location, kind, and size of all trees.
9. Give following elevations: (*a*) outlet elevations of all manholes; (*b*) grade of sewers, streets, and sidewalks; (*c*) bench mark near building site; (*d*) 1-ft. contours over whole tract, if ground is irregular.
10. Show size and kind of buildings on adjacent lots.
11. Show all legal descriptions and certificates usually shown on city plats.

The lettering on such a map is usually confined to single-stroke Gothic, both capital and lower case.

Hydrographic Maps.

The general method of finishing a hydrographic map is similar to other topographic maps. Hydrography has its own particular symbols, shown under the chapter on Conventional Signs. The amount and kind of information shown varies with the use of the map, *i.e.*, a harbor survey should show enough shore-line topography to locate and plan wharves, docks, warehouses, and the location of all streets, railways, and roads along the water front. A navigation chart would show only the shore line and such landmarks as are useful to a pilot in navigation, such as churches, spires, lighthouses, etc. Maps of rivers should show both high- and low-water mark, together with all shore-line topography within this zone. The following information should appear on hydrographic maps:

1. To what datum elevations are referred.
2. Show high-water line as heaviest line on map, and low-water line next heaviest.
3. Letter soundings below datum in black ink, above datum in colored ink. Soundings are given in feet and tenths, the decimal point occupying the exact location of the sounding.
4. Lines of equal depth are interpolated from the soundings and drawn according to symbols shown under Conventional Signs.
5. In navigation charts, the interval for lines of equal depth is 1 fathom or 6 ft. For dredging rivers or harbors the interval for submerged contours is taken as 1, 2, or 3 ft.
6. Conventional signs used for sand, swamps, trees, etc. are the same as on topographic maps.
7. Lighthouses, light ships, buoys, etc., are shown either by conventional sign or lettered on the map.
8. Submerged contours are shown by dash and dot lines, the number of dots between the dashes representing the fathoms of depth. See "fathom lines" under chapter on Conventional Signs. Figure 3 shows a typical sounding chart for harbor maps.

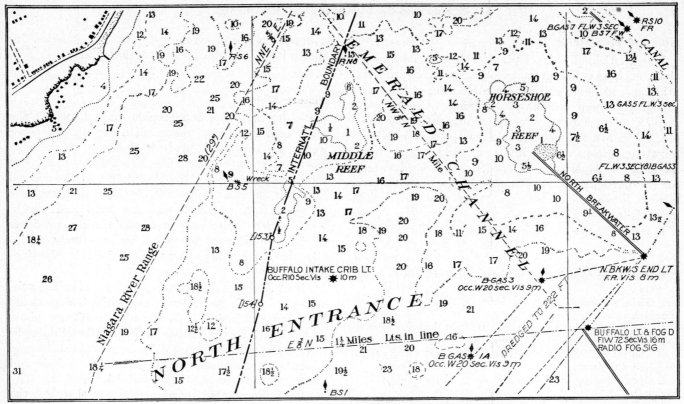

Fig. 3.—Portion of hydrographic map. (*By permission of War Department U. S. A.*)

Golf and Polo Fields.

New and rapidly developing features of landscape work, such as golf courses, polo fields, aviation fields, athletic fields, etc., call for a combination of the true topographic map and the picture map. Clubhouses, stables, trees, shrubs, streams, roads, walks, drives, etc. are shown to large scale in true plan view, while fairways, greens, tees, bunkers, sand traps, and hazards are superposed in special symbols invented for the particular kind of map. Figure 4 shows the layout for a golf course and polo field, with the symbols in most common use.

The requirements for such a map are as follows:

1. Scale large enough to show detail, 1 in. = 50 ft. to 1 in. = 100 ft.
2. Main topographic features shown as near as possible in true plan view.
3. If it is desirable to show relief, use light dotted contours to avoid obscuring important details.
4. The map should have an ornamental appearance as to trees and shrubbery.
5. Open ditches, bunkers, and steep slopes shown by hachures.
6. Fairways marked as to outline and number.
7. Sand traps, tees, and other objects shown by symbols recorded in a legend near the border of the map.

Special plates of symbols for golf and aviation are shown in the chapter on Conventional Signs.

Picture Maps.

Picture maps are topographic maps in which the scale used is relative rather than actual. In other words, no attempt is made to delineate the various features shown on the map to an exact scale, but rather to present an attractive picture of the general layout of the area covered with frequent overemphasis of vegetation features, such as trees and shrubbery. Such maps are not engineering products in the true sense of the word, and are used principally for advertising purposes, real-estate developments, etc.

Figure 5 shows a picture map of a real-estate subdivision. Boulevards, parkways, drives, and shade trees are emphasized. Names of streets are given to fix location, and houses under construction indicated by symbol. Figure 6 is also a real-estate map whose main purpose is to advertise the athletic and amusement features the subdivision offers. The border is illuminated by numerous pen sketches, the artistry of which completely subdues the engineering features of the map.

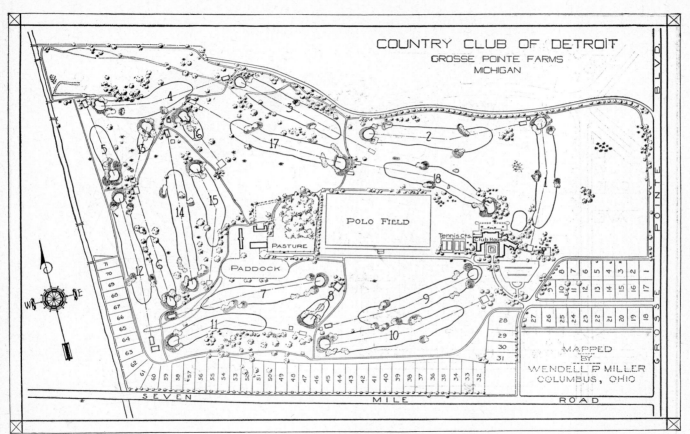

Fig. 4.—Map illustrating golf topography.

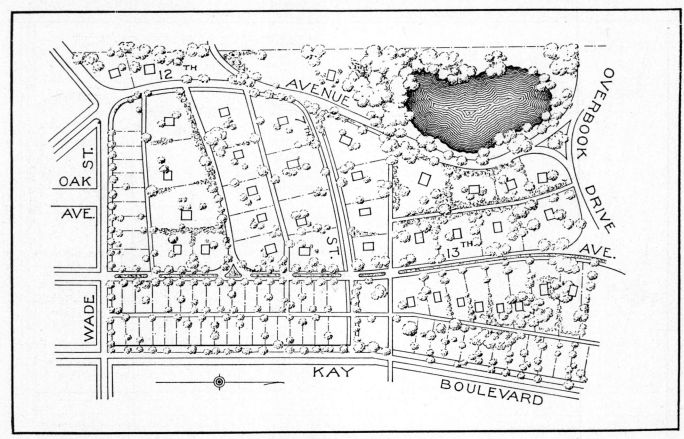

Fig. 5.—Real-estate display map.

Landscape Maps.

If the scale of a topographic map is increased until all features of topography and culture can be plotted accurately to scale, the resulting map is a true plan view. This is a condition seldom attained, yet the nearer a map approaches this ideal condition, the more valuable it becomes as an aid in landscape work.

Such maps are used as a basis of study in landscaping parks, cemeteries, private estates, and residences. Studies are made to select the location for new buildings, to supplement present vegetation with trees, shrubs, and flowers; and to determine possible changes in relief. In order that the above map locations may be reliable, field measurements must be carefully made and the map plotted to a large scale. The scale used ranges from 1 in. = 10 ft. to 1 in. = 50 ft.

On such a map, buildings, roads, walks, manholes, catch basins, fire hydrants, etc., may be plotted to scale in true plan view. Trees can be designated as to size and species, and exact outlines of flower beds and shrubbery shown. Some architects require that the spread of branches for all important trees be shown and a legend placed near the border of the map describing the present state of preservation of every tree.

The scale of the map being large, the contour interval is generally reduced to 1 ft. or 6 in. Existing relief is shown in full contours and proposed changes in dotted contours. The above-described map is a high-grade topographic map made to convey a mass of information to the landscape architect and is not to be confused with landscape *display* maps drawn for the purpose of presenting to a prospective owner the artistic arrangement of buildings, walks, drives, shrubs, and terraces to harmonize with existing topography.

Figure 7 is a reduced copy of a landscape architect's study, made to display the draftsman's idea of a pleasing arrangement of culture, vegetation, and relief for a country estate. Being a display map, buildings are made full and black to give them proper emphasis, trees are rounded, symmetrical, and shaded to produce relief, paths are inked only in open places among the trees to suggest cool shady walks, water lining is shaded near the shore line to make it more prominent, and contours are shown in light dotted lines to interfere as little as possible with the display features. An ornamental north point and an illuminated or shaded title is generally used in combination with such a map, but have been omitted, in this illustration, for lack of space.

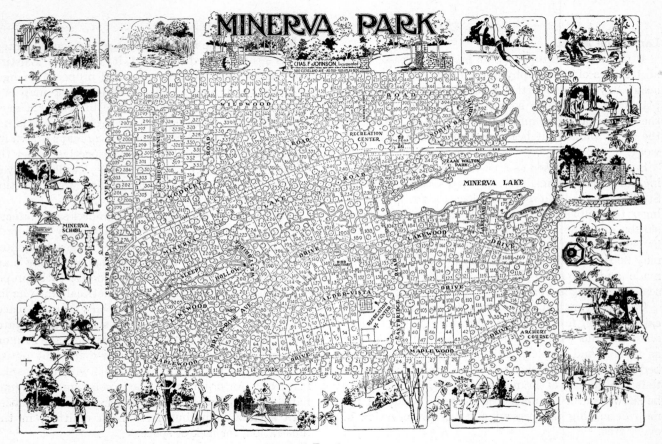

Fig. 6.

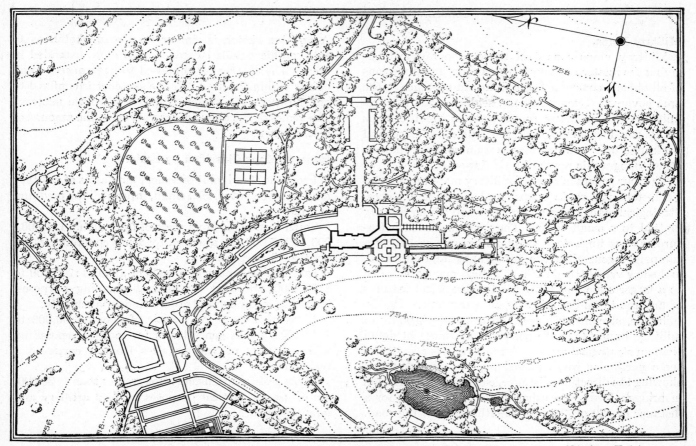

Fig. 7.—Portion of landscape architects' map. (*By permission of Miss Elusina Lazenby, Fine Arts Dept., Ohio State Univ.*)

Military Maps.

Military maps consist of those topographic maps giving the relative distances, elevations, and directions of all objects considered of military importance in the area covered. Such maps are seldom made as carefully or with as many refinements as are the maps previously mentioned, and the amount of freehand work used on a military map is usually larger than on other topographic maps of a similar scale.

The four general classes of military maps are: strategical, tactical, route maps, and position sketches. Strategical maps cover a large area and are necessarily plotted to a small scale. They emphasize such features as are important in planning and executing a campaign, generalize the topography and drainage, and show very little detail. Tactical maps are drawn to a larger scale and show the topographical features of an area to be occupied by an army unit. Roads, railroads, streams, bridges, and other topography of military importance are shown much more in detail than on strategical maps. A route map is a sketch of some particular route passed over by the sketcher, showing the direction, distance, elevations, description of bridges, fords, and other important features for a distance of 300 or 400 yd. on either side of the road.

A position sketch is similar to a road sketch as to topographic details located, but is more extended in area. The sketcher is supposed to have access to all points within the area. If the sketcher is confined to one position, the sketch is called a "place sketch." The scales of the above maps vary with the amount of detail shown.

Military Maps Classified by Scale.

From the standpoint of scale, four classes of military maps may be noted, *viz.*, small, intermediate, medium, and large.

Small maps made to a scale of from about 1:1,000,000 to 1:7,000,000 are used by army commanders for strategical purposes.

Intermediate maps of a scale of about 1:200,000 to 1:500,000 are used for general planning purposes, including troop concentrations and troop supply.

Medium maps using a scale of about 1:50,000 to 1:125,000 are in general use for many purposes by military units of various sizes from a corps to a regiment.

Large-scale maps, usually not greater than 1:20,000, are of use for tactical purposes by field artillery and infantry units.

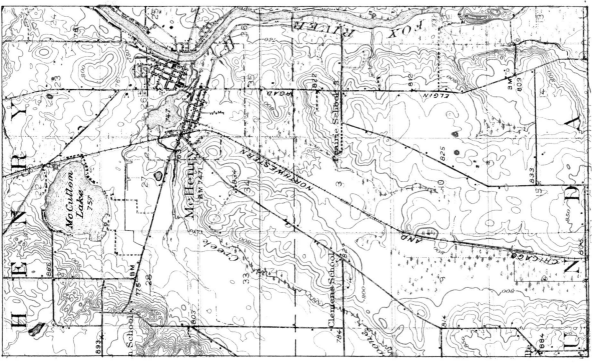

FIG. 8.—Portion of typical contour map of U. S. Geological Survey. Scale approximately 1 in. = 1 mi. (representative fraction 1/62,500).

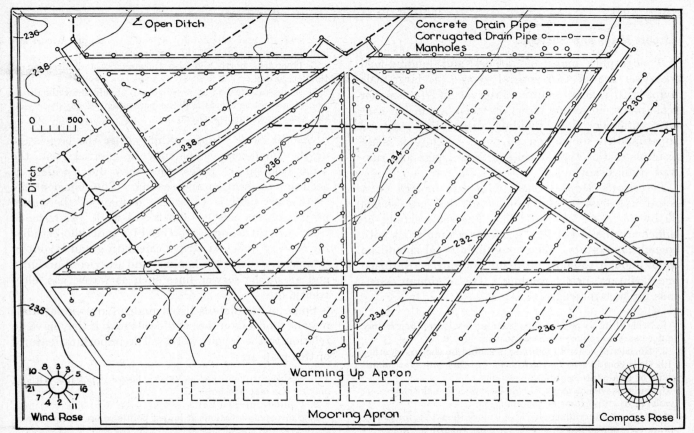

FIG. 9.—Map illustrating airport studies. (Bench marks, horizontal control and declination to be shown on map.)

Airport Maps and Surveys.

Topographic maps of airport and landing-field locations are of two general types: (*a*) large-scale maps with a small contour interval and (*b*) smaller scale maps, all in one piece, on which studies of runway, drainage, and building disposition may be superimposed for design purposes. Figure 9 is an example of this type of map with drainage and runway studies shown.

Large-scale Maps.—These maps are plotted on a coordinate system and usually divided into sheets 30 by 36 in. or 24 by 30 in. The scale should be sufficiently large to allow accurate scaling from both tracing and prints. The contour interval should be small enough to show all details of surface drainage and facilitate surface grading. The minimum essentials for such maps are

1. Scale: 1 in. = 50 ft. unless tract is very large; then it may be necessary to reduce the scale to 1 in. = 100 ft.
2. Horizontal control monuments should be shown, together with their coordinates and azimuths to other adjacent monuments or local state coordinate systems.
3. Sufficient bench marks should be shown to make elevations available to all portions of the map.
4. Contour interval should be not more than 1 ft. for gently rolling terrain and not over 2 ft. for more rugged territory.
5. Culture and topography should be mapped accurately and lightly so that it will not obscure contours.
6. Give true north and also the amount and direction of the declination.
7. A diagram (called a "wind rose") should show the direction of prevailing winds and the percentage of days per year the wind blows in each direction.

Smaller Scale Maps.—While either the large-scale or the small-scale maps may be plotted from the same field-survey notes, they may differ widely in accuracy of location and scaling. The object of the small-scale map is to secure a map of the entire project on one sheet. Black and white prints made from such a tracing can be used for individual studies for runway and taxi strips, calculation and layout of drainage structures, hangar and service buildings, and all public service lines such as water, gas, electricity, etc.

Some of the details listed under Large-scale Maps may be omitted on the small-scale map if they have no relation to the layout work. The important features of this map are:

1. Scale not over 1 in. = 500 ft., larger if possible without making the map too large.
2. Contour interval 2 ft.
3. Show spot elevations at lowest points where **surface drainage** leaves the tract.

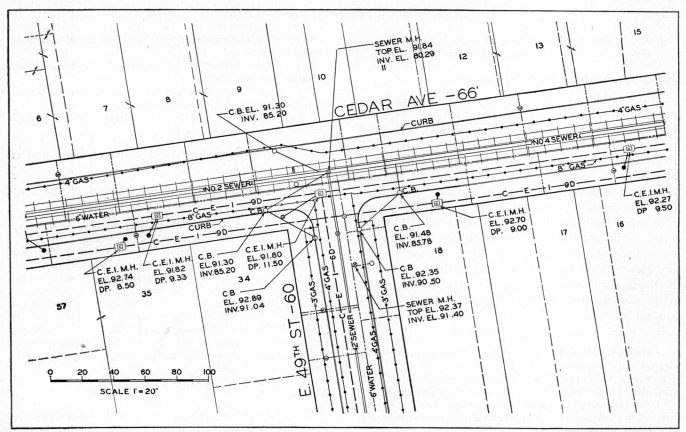

Fig. 10.—Map showing typical underground survey. Coordinate lines to be shown. (*By permission of Cleveland Regional Underground Survey.*)

4. Should have a wind rose to show prevailing winds.
5. Should show true north and compass declination.

Mapping Underground Surveys.

Many public utilities, such as sewers, water lines, gas mains, telephone conduits, etc., are placed beneath the surface of the streets and sidewalks. It is of prime importance to know their exact location with reference to curb and property lines so that repairs and renewals may be made with the least damage to surface structures. In compiling such a map, surface topography is almost entirely omitted except such street and land lines as may be needed to show exact locations. Symbols representing underground structures are drawn with bold heavy lines for emphasis, while surface features are kept in the background by the use of lines much lighter in weight. Figure 10 represents a small section of the Cleveland Underground Survey.

The main features of such a map are:

1. Should be mapped on a coordinate system and tied to the city triangulation survey.
2. Should be plotted to a large scale to avoid crowding of symbols. 1 in. = 20 ft. is the scale most frequently used.
3. Symbols should be boldly executed to produce emphasis.
4. All symbols not recognized as standard should be included in a legend on the map.
5. On all sewer manholes both curb and invert elevations should be given.
6. On other public utility manholes both cover and bottom elevations are recorded.
7. Curb and lot lines should be inked lightly and be less prominent than underground features.
8. Lettering should be vertical commercial Gothic for both capitals and lower-case letters.

Classes of Topographic Maps—Based on Method of Delineation.

From the standpoint of preparation of the map itself, two classes of maps may be defined, *viz.*:

1. Plain topographic maps consisting of those drawings on which all symbols are shown in black and white.
2. Colored topographic maps on which the symbols are expressed by the use of various color combinations and tints. The colors used may be either colored inks or water colors. Where colors are used in the preparation of a topographic drawing, the four general classes of topographic features are commonly given the following characteristic colors: Culture is shown in black, or at least outlined in black, relief in brown, hydrographic features in blue, vegetation in either green or black, and natural vegetation, such as trees, forests, underbrush, meadows, etc., in green.

Map Scales.

The scale of a map may be defined as the constant ratio between the actual dimensions of objects and distances on the ground and the representation of these dimensions and distances on the map. The objects on a topographic map being shown in plan view, the dimensions on a map to which its scale applies will be horizontal distances only. Vertical distances, however, may in effect be shown to scale by means of such devices as the contour interval. The accuracy with which vertical dimensions or elevations can be taken from a map will depend upon the accuracy used in the horizontal scaling or plotting of the contours themselves.

It should be borne in mind that on many topographic maps only the controlling distances and most important objects are shown to scale, the general background of such maps, consisting of symbols representing forests, trees, meadows, cultivated land, etc., not being shown to an exact scale. On such maps, however, the *relative* importance of the various objects shown should not be lost sight of, and the less important features should not be emphasized in size and weight of line at the expense of the more important features. In the case of landscape and similar large-scale drawings, practically every item on the map, including trees, houses, paths, roads, and other features, will be shown to an exact scale, because all have an equally important bearing on the ultimate object in view.

The scale is one of the most important items to be taken into consideration in the construction of a topographic drawing, and the choice of scale will depend upon the purpose for which the map is to be used and the accuracy desired, which in turn will depend upon the accuracy of the field work upon which the map is based. Certainly, there ought to be a definite relationship between the accuracy of the field work and the accuracy of the map work. The best practice seems to indicate that the relation between survey accuracy and map scale should be such that the average error in the horizontal position of a point should not exceed twice the smallest plotting unit, and the average vertical error should not exceed one-half the contour interval.

Generally speaking, the scale of the map should be such as to keep plotting errors within the same limit as the errors of the field work upon which the map is based. Since plotting and field errors are not likely to compensate, it is best for the draftsman to keep the plotting error considerably closer than the limiting error in field work. Two principal factors govern the relation between scale of map and contour interval:

1. The character of the relief—whether mountainous, rolling, or flat.
2. The accuracy with which it is desired to read elevations from the map.

For rugged or mountainous country, a map scale must be selected which will allow a clean-cut delineation of each contour and avoid obscuring other important features of topography with a mass of closely drawn contours. On the other hand, the smaller the contour interval, the closer elevations may be read from the map.

The following table gives general practice for usual conditions:[1]

Range of Scale	Slope of Ground	Interval in Feet
1 in. = 10 to 100 ft.	Flat	0.5 or 1
	Rolling	1 or 2
	Hilly	2 or 5
1 in. = 100 to 1,000 ft.	Flat	1, 2 or 5
	Rolling	2 or 5
	Hilly	5 or 10
1 in. = 1,000 to 10,000 ft.	Flat	2, 5 or 10
	Rolling	10 or 20
	Hilly	20 or 50
	Mountainous	50, 100 or 200

[1] Table reprinted with permission of Davis, Foote, and Rayner.

Methods of Indicating Scales.

The scale of a map is commonly indicated on the map itself in one of three ways:

1. By a "representative fraction," as 1/1,000.
2. By an expressed ratio between a unit of scale on the map and a different unit of measurement on the ground, as, 1 in. = 100 ft.
3. By graphic representation.

In many instances, more than one method is used to express the scale of a map, as on the quadrangle sheets of the U. S. Geological Survey.

By "representative fraction" is meant a fraction with a numerator of unity and a denominator a multiple of 10, which expresses the ratio between a given unit distance on the map and the same distance expressed in the same kind of units on the ground. For example, a representative fraction of 1/1,000 would indicate that 1 in. on the map is equivalent to 1,000 in. on the ground. Expressed in the form of an equation:

Let S = a plotted distance on the map
L = the corresponding distance on the ground in the same unit as S
F = some multiple of 10

Then, $1/F = S/L$, in which $1/F =$ there presentative fraction. Comparing the methods of 1 and 2, a scale of 1 in. = 200 ft. is equivalent to a representative fraction of $\dfrac{1}{200 \times 12}$, or $1/2,400$.

It is an easy matter by using the expression $1/F = S/L$ to convert actual distances on the ground into equivalent plotted distances. For example, a distance of 586 ft. measured on the ground is equivalent to

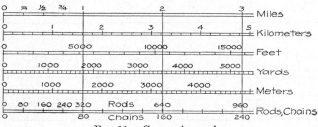

Fig. 11.—Conversion scale.

2.93 in. on a map with a representative fraction of $1/2,400$, or $1/2,400$ equals $\dfrac{S}{586 \times 12}$, from which $S =$ 2.93 in.

By making use of the above expression, the representative fraction of a map may be obtained by comparing a measured distance on the ground with its equivalent plotted distance, provided proper allowance can be made for shrinkage or expansion in the map itself.

Under ordinary conditions, the average draftsman cannot plot a point with the naked eye closer than $1/100$ in. Accepting this or a similar limit as correct, it is an easy matter to compute the limits within which distances can be exactly laid off or scaled from a map of a given scale. For example, let the representative fraction of a certain map be $1/3,000$, or 1 in. = 250 ft., with the draftsman capable of plotting to $1/100$ in. No distance could be plotted on, or accurately scaled from, this map closer than a ground distance of $3,000 \times 1/100 = 30$ in., or $2\frac{1}{2}$ ft. It should be noted, however, that this limit affects all distances equally within the length of the scaling rule used, *i.e.*, using a 12-in. scale, a distance of 10 in. on the map may be laid off with the same accuracy as a distance of 1 in.

If allowable plotting error on a certain map is limited to 5 ft., the corresponding map scale may be found as follows:

From the relation

$$\frac{1}{F} = \frac{S}{L}, \ L = FS$$

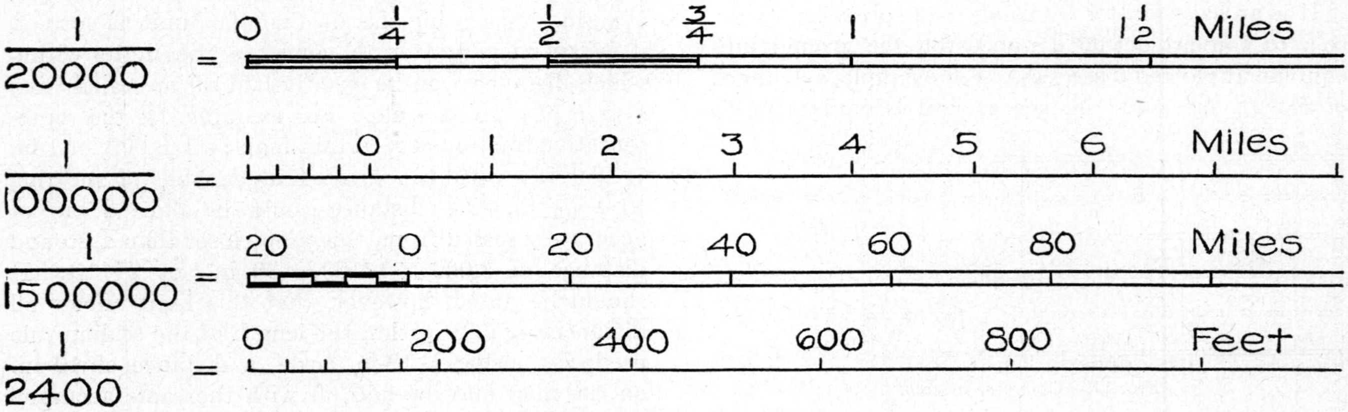

FIG. 12.—Types of bar scales

or

$$L = 100 \times 12 \times 5 = 6,000$$

i.e., a scale of 1/6,000 would keep the plotting error within the 5-ft. limit.

In the use of maps from which areas are to be scaled, such as right-of-way maps, reservoir sites, park sites, etc., the above methods of estimating the

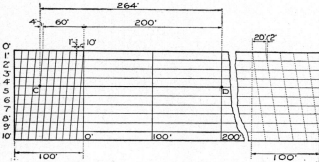

Fig. 12a.—Proportional scale (diagonal scale).

limit of scaling error must be modified to the extent that the error of scaling must be taken into consideration more than once. For example, in the case of a map 10 in. square, representing an area of 1,000,000 sq. ft. and plotted to a scale of 1/1,200 or 1 in. = 100 ft.: Based on a scaling accuracy of $\frac{1}{25}$ in., the area could be scaled from it with an error of 0.8 per cent, whereas the distances are scaled with an error of only 0.4 per cent, the percentage of error in scaling the area being double that in scaling the sides.

Occasionally the draftsman may be called upon to construct a special scale. Figures 11, 12, and 12a show types of such scales. Figures 11 and 12 are self-explanatory. Figure 12a shows a proportional or diagonal scale also called a "plotting scale." This scale can be constructed to fit any representative fraction. For example, to construct a scale for 1:1200 to read up to 1,000 ft.: The length of the scale will be $\frac{1,000 \times 12}{1,200} = 10$ in. This length may then be divided into 10 equal parts, as shown, each of which will measure 100 ft. The two end rectangles may then be further subdivided by diagonal lines, as shown, to read in smaller units.

Tests for Accuracy of a Map.

The best test of the accuracy of a topographic map is a comparison between scaled horizontal and vertical dimensions on the map and the same dimensions on the ground. If it is desired to apply such a check to a map, one or more test lines should be run on the ground covered by the map in question, the lines being run on a definite azimuth and then plotted to azimuth on the map. A profile taken along this line in the field should, when plotted, compare favorably with a profile of the same line plotted from elevations and distances scaled from the map. In a similar manner, a horizontal check may be made of the same line by comparing the plotted positions of all street crossings, fence-line crossings, house locations, etc., on the map, with the location of the same points as located by the check work in the field.

CHAPTER II

CONVENTIONAL SIGNS

Only a limited amount of detail could be shown per square inch on a topographic map, if all features were represented to the same scale. For this reason symbols or signs representing the different features, such as culture, relief, and vegetation are used.

These symbols should be of such form as to be rapidly and easily made, readily understood, and pleasing in appearance.

Symbols used for the commonest features, such as vegetation, woods, cleared land, water and streams, have become a standard in the engineering profession, while special symbols of culture—crops, parks, etc.—may vary with the taste and skill of the draftsman.

A topographic map corresponds to the plan view in orthographic projection. The symbol should resemble the plan view of the object which it represents.

Many of the symbols, however, such as crops, telephone poles, etc., are frequently made in elevation, because this view is more pleasing in appearance, and more easily understood.

Shading of Symbols.

The effect of light and shade is often used to produce relief in the individual symbols themselves.

The light ray occupies a position in space such that its plan projection makes 45 deg. with a horizontal line. The source of light is regarded as being above and to the left of the object. Objects extending above the horizontal plane are shaded on the side farthest from the source of light, while objects below the horizontal plane are shaded on the side nearest. Positions of shadows when used are similarly determined. Shading is used on landscape and large-scale topographic maps, but it is generally omitted on small-scale drawings.

Composition.

The combining of the various symbols into a neat and intelligible map is as much a matter of composition as the writing of a beautiful sonnet, or the arranging of the notes of the musical scale into a pleasing symphony. Clearness should be combined with tasteful execution. Symbols should be just large enough to be plainly and easily read. The purpose for which the map is made should largely govern the composition, *viz.:* A map showing the proposed method of landscaping a private estate should show trees, shrubbery, flowerbeds, walks, and drives boldly, and shaded to produce relief; while contours and other relief symbols should be inked lightly.

A map whose main purpose is to show relief should have the contours inked as the prominent feature, while trees, shrubbery, etc., necessary to be shown, should be inked lightly so as not to obscure the relief.

A good rule to follow is to select the most important feature to ink first. (This should be the first feature seen, when viewing the finished map.) Then ink the remaining features in the order of their importance, toning down the weight of the lines in proportion.

The least important symbol, such as grass or cleared land, should be used to tone up the map, *i.e.*, to bring the map to a soft, even tone. This is done by filling white open places with the symbol until the map, when held at a correct reading distance, appears to have an even undertone, while the important features seem to stand out above the general background. To assist in attaining this end, a detailed discussion of the various symbols, together with the faults most commonly committed in drawing them, is given a prominent place in presenting the above subject to the student.

Choice and Use of Pens.

The pens ordinarily used are Gillott's 170, 303, and 404. For the finer work, 290 or 291 mapping pens, or any good crow-quill pen, should be used. The beginner should experiment with the above pens and select the pen he can use best.

A pen should be so handled that it appears to move easily and steadily on the paper, with little more than its own weight for bearing. A person having the ability to use a pen in this manner can successfully use a limber pen such as 290 or 291 for sketching or water lining. A draftsman not able to develop so light a touch should use a crow-quill or Gillott

170 for similar work. The beginner should execute such symbols as grass, trees, underbrush, etc., with Gillott 170 pen and change to the 290 and 291 mapping pens as he develops skill.

Rules for Making Symbols.

1. Do not carry an excess of ink on your pen or allow it to become entirely dry before refilling.
2. Do not allow ink to become dry and encrusted on your pen.
3. Do not use same pen for black and colored inks.
4. Do not use a pen after the nibs have become sprung.
5. Try the pen on margin of drawing before using.
6. Clean dirty pens with ammonia.
7. Thick ink can be thinned by adding a few drops of ammonia.
8. Use guide lines for all letters and figures.
9. Keep paper clean and avoid erasures.
10. Avoid awkward positions when sketching or lettering.
11. Block out your work very lightly in pencil, as heavy pencil work makes it very difficult to judge shading to be done in ink.
12. Learn to make the symbols uniformly and well before you attempt to place them on the map.

Symbols and Signs.

For convenient reference, many authorities group the various signs and classify them as *land, land and water,* and *miscellaneous*. It is not the intention of the authors to follow any given classification in presenting the subject of topographic symbols to the student, but rather to group them for ready reference according to their various uses, such as military, railroad, oil and gas, works and structures, etc.

The various symbols appear on attached plates lettered to indicate each symbol. Many of these symbols used are purely mechanical in outline and can be executed by any one skilled in mechanical drafting, and need no further explanation. Symbols showing relief, vegetation, hydrography, etc., require considerable skill in freehand drafting, and will be discussed in detail. *Grass or cleared land* is shown by groups of five to seven dots or dashes arranged symmetrically about a center line, and distributed evenly over the area to be covered. The group or tuft of grass should have its base even and parallel to the bottom border of the map. The sequence of strokes and development of the symbol are shown in enlarged form in Fig. 1, Plate 1. The effect desired, when viewed at a short distance, is that of a flat tint. This may be accomplished as follows:

1. Draw light pencil guide lines at irregular intervals, and parallel to bottom border, over the area to be covered.
2. Sketch tufts of grass at irregular intervals, but well separated, over the entire area.

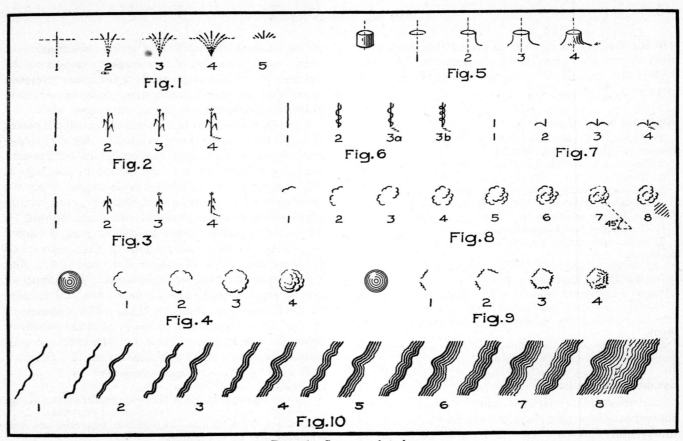

PLATE 1.—Sequence of strokes.

3. Make an effort to produce an even tint by proper distribution of tufts. At the same time avoid placing the symbol in rows.

4. View your work from a distance of 14 to 20 in., and bring the area to blend with the remainder of the map by inserting smaller tufts or rows of dots in spots apparently light.

The usual faults are as follows:

1. Symbol too large or out of scale with map.
2. Base of tufts not parallel to bottom border.
3. Placing the tufts in rows.
4. Failure of grass plot to blend with adjacent symbols or portions of map.
5. Not keeping the strokes separate and distinct.
6. Strokes in tuft too widely separated.

Recently cleared land is often represented by replacing a few of the tufts of grass by a symbol showing a stump in perspective. The symbol and sequence of the strokes are shown on Fig. 5, Plate 1.

Indian Corn.—The symbol used to represent Indian corn is a single stalk of corn in elevation. The symbol is placed in rows at a distance apart required by the relative shading of the map. Care must be taken not to make the symbol too large, and uniformity is of prime importance. The symbol is difficult of execution, and is used where a designation of crops is essential. For large-scale or landscape work, the symbol shown in Fig. 2, Plate 1, should be used. For smaller-scale topography, use symbol as shown in Fig. 3, Plate 1. The general faults are: symbols too large, rows too close together, symbols not vertical, and lack of uniformity.

Deciduous Trees.—The tree symbol consists of small curved outlines representing the tree in plan. An attempt is made to give the tree relief, *i.e.*, an apparent dimension, perpendicular to the paper. This is accomplished by simulating the shading of a sphere in perspective as shown in Fig. 4, Plate 1.

Care should be taken to keep the tree rounded in appearance, shading it on the lower right-hand corner, and leaving a white spot or high light in the upper left-hand corner. Some draftsmen accomplish this effect by variation in weight of lines, while others get a similar result by sketch shading.

Oak Trees.—These trees are sometimes designated by a symbol similar in outline and shading to deciduous trees, but having the convex portions of the strokes facing inwards, as shown in Fig. 9, Plate 1.

Orchard.—Orchards are represented by using the deciduous tree symbol as described above, laid out in regular rectangular spacing, each tree having an individual shadow. The size of squares showing the tree spacing is governed by the scale of the map, and should be laid off with the rows parallel in one direction with the bottom border. The shadow of the tree is sketched about an axis passing through the center of the tree and making 45 deg. with the bottom border, as shown in Fig. 8, Plate 1. The group symbol is shown on Plate 7, marked "*Orchard.*" The shadow is made by a series of parallel freehand strokes at right angles to the axis of the shadow. The shadow should be started with a short line nearly touching the tree, gradually increasing the length of the line until the shadow is the same width as the tree; then gradually decrease the length of line, until the shadow is symmetrical. The common faults are as follows:

1. Angularity of outline.
2. Failure to produce relief by shading.
3. Shadows not parallel and often too large.

Evergreens.—Evergreen trees are represented in plan by a five-pointed star, as illustrated in group formation and by a single magnified symbol on Plate 2. This symbol is best made with a limber, flexible pen, 290 or 291. Place a pencil dot marking the center of each tree. Make one point of the star for each tree by starting the stroke with considerable pressure of the pen, gradually diminishing the pressure as the stroke is drawn. This will produce a line increasing in width from the point toward the center of the tree. Turn the map through one-fifth of a circle and repeat until all five points are drawn. Care must be taken not to allow the strokes to touch one another as they converge toward the center. Some draftsmen execute the symbol by drawing five radiating lines from a dot in the center. This symbol is not pleasing unless all lines are kept radial, which renders its execution very difficult.

Woods.—A combination of two symbols, *i.e.*, of evergreens and deciduous trees, is used to represent woods. This sign is somewhat similar to that for orchards, except that the spacing and size of trees are irregular and the shadows are omitted. Angularity in outline and regularity in distribution are to be avoided. If clumps of trees are drawn they should be kept rounded in outline, avoiding the tendency to allow the individual strokes to cross and form loops. If woods are interspersed with brush, they should be shown as on Plate 7.

Vineyard.—This symbol is made by a straight vertical stroke representing a stake or pole, and a downward reverse-curve stroke to picture the vine encircling the stake. Two methods of making this stroke are shown as 3*a* and 3*b* on Fig. 6, Plate 1.

Figure 3a is to be preferred for its simplicity and ease of execution.

Tobacco.—This symbol is shown in Fig. 7, Plate 1, and in group symbol on Plate 7, and is one of the most difficult to draw and keep uniform in appearance. It is best executed by the use of a fine flexible pen which will permit shading the middle portion of the stroke by increased pressure. The most common faults are: not keeping the two sides symmetrical; and connecting all three strokes.

Cultivated Land.—Where it is not desirable to designate particular crops, a symbol is used to represent cultivated land, as shown on Plate 7. A series of parallel lines are ruled, covering the space to be mapped. This is done with triangle and right-line pen. The dots are made with a mapping pen in the following manner:

Turn the paper until the draftsman is looking lengthwise of the ruled lines. The dots are then made by working from the far end of the lines toward the draftsman. Make the dots by lifting the pen vertically from the paper and space them slightly farther apart than the line spacing. Another system much in vogue is to draw alternate dot and dash lines with the right-line pen. While this reduces the amount of freehand work, it is almost impossible to keep the work uniform and maintain a reasonable speed.

Plate 2 shows the common vegetation symbols magnified several diameters to give the student a better opportunity to study the component strokes making up the symbol. The student should experiment with the sequence of strokes making up the various symbols until he arrives at a method of making the symbol which is both pleasing in appearance and easily executed.

Rice.—The symbol for rice consists of a single tuft, similar to the grass symbol, with two or three horizontal strokes parallel to the base of the tuft.

It may be shown in mass, either with or without a series of irregular areas bounded by dotted lines to represent the embankments or paddies used in flooding.

To represent rice cultivation with paddocks:

1. Divide the area into subdivisions of irregular size and shape bounded by lines slightly curved. The size of the subdivisions vary from $\frac{1}{2}$ to $\frac{1}{4}$ in. square according to the scale used. The conventional color for the paddocks is brown, although black is often used.

2. In penciling the irregular areas, avoid arranging the area in regular tiers; using too many triangular

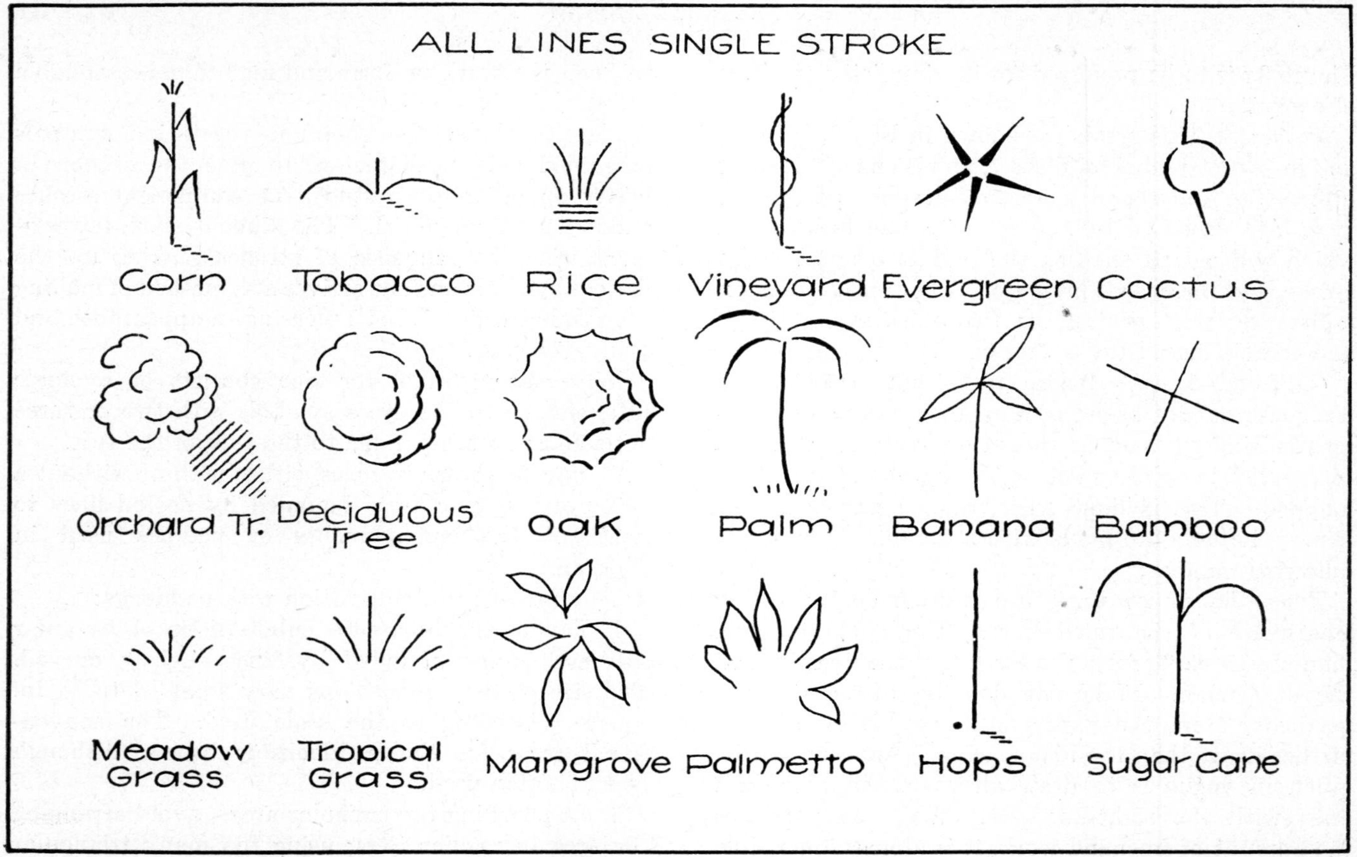

PLATE 2.—Vegetation symbols magnified.

or five-sided figures to break up regularity; and the use of markedly curved boundaries.

3. Cover the whole area with the rice symbol, using the above symbol to even up the texture of the map.

4. The texture should be as coarse as is consistent with good appearance. Fine texture increases the labor required in execution.

5. The paddocks are often omitted from the symbol unless the rice is shown in green; the paddocks are then inked in brown.

Sugar Cane.—The symbol consists of a vertical stem representing the stalk and two curved drooping lines, attached to the stalk at about two-thirds its height, to represent the blades. Being a cultivated crop, it is planted in rows similar to corn. To execute the symbol:

1. Cover the area with ruled squares about twice the dimension of the symbol used, the sides of the squares to be perpendicular and parallel, respectively, to the bottom border of the map.

2. Draw the stems or stalks to terminate at the line intersections.

3. Draw the blades at approximately two-thirds the height of the stem, allowing just enough variation to avoid monotony.

4. The spacing of symbols is varied to suit the area covered and the scale of the drawing.

Wheat.—No conventional sign has been adopted for wheat, although it is the most important of the cereals. The symbol used is suggested in Stuart's, "Topographical Drawing," and is purely conventional. It consists of a stem, vertical for about two-thirds its height, and slightly curved to the right for the upper one-third. Two curved strokes are placed at the right of the upper third joining the stem at an acute angle about where its upper curved portion begins. The symbol is not executed in rows, as rows are not apparent in the growing crop.

Oats.—The symbol suggested for oats is similar to that used for wheat, with the exception that three small vertical strokes are used at the right and under the upper third of the stem stroke. The distribution of the symbol should conform to the same rules as for wheat, as the crops are very similar.

Hops.—The symbol is executed by a straight vertical stroke similar in size and spacing to the vineyard symbol already described. A small dot is placed to the left and near the base of the vertical stroke. The symbol represents a stake or pole driven into the ground. A simulation of the vine, as in the vineyard symdol, is omitted to avoid confusion.

Spacing and distribution over the map area are the same as are used for vineyards.

Cotton.—The symbol used is purely conventional and consists of a small freehand circle with a dot in its center. The dot should be made first, as it is difficult to place dots in the center of small circles rapidly. Dots badly out of center spoil the looks of the symbol. The symbol should be in rows, in one direction only, as cotton is drilled or planted in rows and not checkered as is the corn symbol. The rows should be parallel to the longest boundary line, if it is a straight line; otherwise, the direction of the lines may be chosen arbitrarily.

Cactus.—The symbol for cactus consists of a small freehand circle with one, two, or three radial spines extending outward from the circle a distance approximately equal to the radius of the circle. The symbol should vary in size and distribution with the area covered and the scale of the map, but should never be large enough to make its execution slow and difficult.

Bananas.—The symbol consists of an upright stem and three pointed leaves. Variation from the monotony of this symbol may be secured by inclination of the stem and variation in direction of the component leaves. To make the symbol in mass, space the individual symbols from two to two and one-half times their diameter apart, filling in any noticeable white spaces by symbols reduced in size.

Palms.—The symbol for palms consists of from two to four curved freehand strokes to represent the foliage, a single upright stroke to represent the trunk, and two to three vertical strokes on each side of the base of the trunk to balance the symbol. The strokes at the bottom resemble the component strokes in a tuft of grass except that they should be vertical. To show palms in mass, group the individual symbols irregularly and space to match the tone or texture of the adjacent parts of the map. As in other symbols irregularly spaced, any lack of uniformity may be corrected by use of the symbol in reduced size.

Bamboo.—This symbol is a small cross, executed freehand. The strokes must be mutually perpendicular and bisect each other. Care must be taken not to make the strokes too long and to avoid monotony. This symbol being purely mechanical, almost the only variations possible are rotation of the symbol in azimuth, and variation in size and spacing

Willows.—The symbol for willows consists of a short vertical stroke representing the trunk of the tree and four branches irregularly arranged in the upper half of the stem. A small horizontal shade line is generally drawn near the base to balance the symbol. Willows in mass are shown by an irregular grouping of the individual symbol. Variation may be secured by curving the upper third of the stem stroke either to the right or the left, but in all cases the lower third of the above stroke must be vertical.

Mangrove.—This symbol consists of from two to six slender pointed leaves on an irregular branching stem. To execute the symbol in mass, draw the larger six-leaved symbols irregularly spaced at about twice their breadth over the entire area. Harmonize with the adjacent parts of the map by inserting smaller symbols in the intervening spaces.

Palmetto.—In common with other tree symbols, palmetto is represented in plan view by a cluster of from four to seven long, pointed leaves radiating from a common center. The leaves must be kept radial or much of the effectiveness of the symbol is lost. Palmetto in mass is shown by irregular grouping of the symbol two to three times its diameter apart, the intervening spaces being filled with partial symbols.

Kelp.—This symbol consists of an irregular curved line breaking into numerous subdivisions and representing a coarse vining seaweed. It is somewhat similar in appearance to small-scale mapping of water courses in badly eroded terrain. The symbol in mass is shown by filling the intervening spaces between larger symbols with smaller or partial symbols until the proper tone is secured.

Symbols Showing Character of Soils.

Sand and Gravel.—The symbol for sand is made by an assemblage of fine dots having no regular arrangement. To represent gravel, larger and heavier dots are scattered throughout the sand symbol. Loose stone or very coarse gravel is shown by sketching small circles or triangles at irregular intervals throughout the sand and gravel symbol. The dots should be round and not have the appearance of commas or short dashes. This can be avoided by carrying a normal amount of ink on the pen and lifting the pen vertically from the paper. The heavier dots for gravel should be made by changing the weight of pen and not by a variation of pressure on it by the draftsman. Sand bars along streams should be shown by decreasing the weight and increasing the spacing of the dots in proportion to their distance from the shore line. This gives the symbol the appearance of a shaded band having its blackest border at the shore line. The top of sand dunes should be outlined by a triple row of dots closely spaced. The spacing of dots is gradually increased from this band outward, until the lightest tint or shade desired on the map is secured. The symbol for the remainder of the area should then be kept uniform.

Mud Flats.—This symbol is represented as shown on Plate 3. It is a combination of irregular areas shaded by horizontal lines and separated by a network of irregular white lines representing the sun cracks. Some draftsmen prefer to draw the shade line with a right-line pen. This method is slow and laborious and tends to give the symbol a stiff, rigid appearance when compared to the adjacent freehand symbols. To ink a mud flat, proceed as follows:

1. Draw the sun cracks with a dull soft pencil so that they can be easily erased.
2. Shade the areas between sun checks by horizontal freehand lines, taking care that the ends of the lines do not touch the penciled sun checks.
3. The pencil work may then be erased and the drawing will have the proper appearance. Gillott's 170 is the best pen for this purpose.

Alkaline flats are shown on Plate 4 as follows:

1. Outline the irregular space to be shown with a soft pencil.
2. Use the sand symbol, following the same rules for gradation in inking the outline.
3. Erase pencil and correct unevenness of tint.

Saline beds are shown in much the same manner, except that the outline is inked by a series of dash lines made vertical to the bottom border, as shown on Plate 4.

Ponds and Small Lakes.—Intermittent ponds are shown by cross-sectioning the areas inside the border with a right-line pen, varying the weight and spacing of the lines to suit the map in question. Salt ponds are shown by filling the area enclosed by the shore line, with a symbol resembling sand, except that the dots are larger and more widely distributed. Larger or permanent lakes are shown by water lining.

Marsh.—Salt marsh is shown by a combination of the symbols for intermittent ponds and cleared land. The former should be ruled first with a right-line pen and the symbol for grass scattered irregularly over the area. Fresh-water marsh is illustrated in the same manner except that long dash lines are used to replace the full lines in the salt-marsh symbol. Both symbols should be made parallel to the bottom border.

Brooks and Small Streams.—These are shown by an irregular line following the course of the stream, fine at its source, and increasing slightly in breadth to the mouth. These lines are more easily drawn by the beginner from the mouth toward the source, as it is easier to diminish the pressure on the pen than to increase it. Since most streams have their source in rugged topography, the portion nearest the source is very irregular and the draftsman should avoid giving his stream lines a snaky, whip-lash appearance.

Rivers and Large Bodies of Water.—The space between the shore lines is filled with parallels called "water lines." This is one of the most difficult symbols to execute properly, and a failure to do so will absolutely mar the map. Draw the first water line at a distance from the shore line equal to the width of the line and following every projection and indentation of the shore line carefully. Draw the next water line parallel to the first and at a distance away slightly greater (about one and one-eighth times) than the space between the first line and the shore. This operation should be repeated, working from both shore lines toward the center, gradually increasing the spacing and decreasing the weight of line. At no place should an abrupt change in spacing be visible. In order that all water lining on the map may appear uniform, it is well to draw all the water lines which are adjacent to the shore lines, throughout the map, and then draw all the second lines; and so on. This will not only ensure uniformity of spacing and weight of line, but will also cause the final lines to meet in the center of the stream, and proper junctions corresponding to the varying width will be formed. The sequence of strokes and spacing is shown on Plate 1, Fig. 10.

In drawing a water line, the preceding one should be at the left hand, as shown in Fig. 13. The line

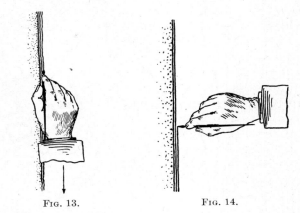

Fig. 13.   Fig. 14.

should be drawn toward the body and the paper so disposed as to facilitate this operation. To preserve the proper interval, keep the eye on the interval, rather than on the line. Figure 14 shows the most common error in position assumed by the student.

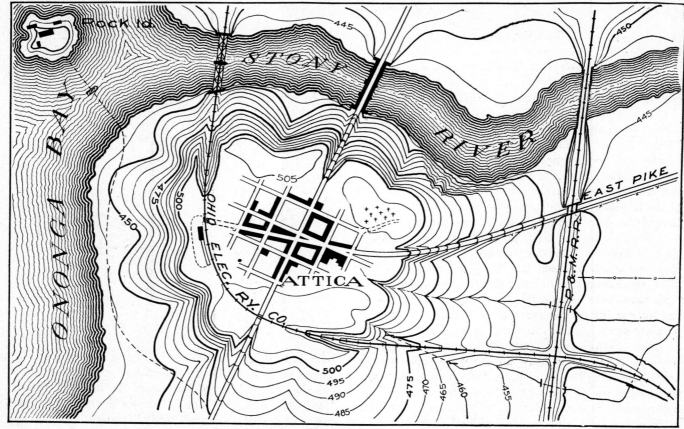

Fig. 15.—Relief shown by contours.

For rapids or torrents, the water lines are broken into short, curved lengths as shown in enlarged view on Plate 5. Whirlpools are given spiral shapes, still preserving, however, the local gradation of weight and intervals, as shown on Plate 4.

On shore-line maps, the water lining is omitted, and sounding locations are shown by figures, to convey the depth in feet. Subaqueous contours or lines of equal-fathom depth are often drawn on these maps as dotted lines, continuous dot lines being 1 fathom depth; dots separated in pairs, 2 fathoms; dots grouped in threes, 3 fathoms; etc. (see Plate 14).

Surface Forms of Ground.—There are two general methods of representing hills, mountains, and surface undulations:

1. By contours.
2. By some form of hill shading.

The first method is most commonly used. It has the advantages of simplicity and speed, and does not obscure the topographic symbols. The second method represents the relief more forcefully to the eye, and if the shading scale is light, does not seriously hide the topography. Hill shading, if properly executed, calls for considerable skill in freehand work and is quite a tax on the draftsman's artistic ability. It is therefore little used on modern topographic maps except to delineate such forms as cuts, fills, eroded banks, rock cliffs, etc., where the inclination of the ground is so great as to cause the contours to run together. Figure 15 shows an area with the relief expressed by means of contours. Figure 16 shows the same area with the relief expressed by hachures.

Drawing Contours.—Contours may be drawn with either the freehand, right-line, or contour pen, depending on the skill of the draftsman. The common faults in the use of the freehand pen are:

1. Too much or two little ink on the pen.
2. Drawing from an awkward position, *i.e.*, pushing the pen sidewise or away from the body instead of turning the paper or body until the draftsman is working toward himself.
3. Failure to start at exact place where inking was stopped to refill pen.
4. Variation in pressure on pen, thus giving a line a variable thickness.

The use of the right-line pen largely eliminates objections 3 and 4, as it will carry enough ink to complete a contour, and does not vary in weight of line.

To use the right-line pen for contour work, grasp the pen slightly below its center between the thumb and first finger, hold it in a vertical position, and draw from left to right, turning the pen about a

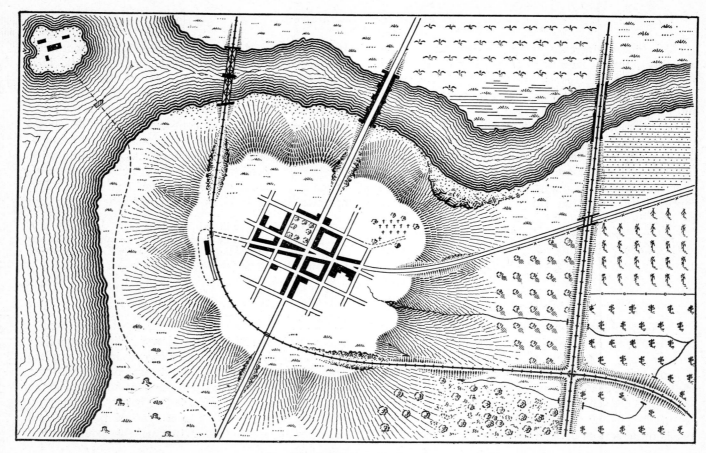

Fig. 16.—Relief shown by hachures.

vertical axis by rolling it between the fingers so that the nibs of the pen are always traveling lengthwise of the contour. With a little practice, the beginner can do very creditable work with this pen.

The contour pen is similar to a right-line pen except that it is fitted with a swivel near the bottom, allowing it to turn in any direction. It is also bent so that the nibs trail the axis of the handle about ½ in. when in use. These two features allow the inking portion of the pen to follow freely the motion of the draftsman's hand in tracing a contour. The correct position for using such a pen is shown in Fig. 17. The draftsman should ink all heavy or accented contours first to ensure uniformity. The intermediate contours can then be inked without the necessity of frequently changing the pen for the variation in weight of line.

Hill Shading.—This method of representing relief has been highly developed by the military engineers of continental Europe, and some of the maps thus produced are wonderful specimens of topographic art. Many systems of vertical, horizontal, and oblique illumination have been developed and are treated at great length and detail in some of the older texts on topographic drawing. This method is too slow and laborious for modern engineering practice, and has largely been discarded in favor of the contour map.

Certain forms of relief, such as cliffs, eroded banks, etc., are not easily shown by contours. A text on topographic drawing should contain a few illustrations and instructions on vertical illumination, by use of hachures.

The shading is accomplished by short strokes of the pen, called "hachures," extending from one

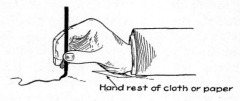

Fig. 17.

contour to the next and following the line of steepest descent. They also break intervals at the contour so as to show its position after the pencil work is erased. The spacing and weight of hachures is governed by some prearranged scale of shading to signify relative steepness of terrain. Usually slopes of 45 deg. or over are shown solid black, and level or horizontal ground is shown white. Intermediate slopes between 0 and 45 deg. are laid out on a shading scale which the draftsman keeps at hand while working on the map.

The beginner will do well to observe the following suggestions, illustrated on Plate 6, until he has gained considerable proficiency:

1. Draw the contours representing the relief as in Fig. 1, Plate 6.

2. If the contours have too great a horizontal spacing to ink the hachures with one stroke of the pen, auxilliary contours should be drawn, as in Fig. 2.

3. Guide lines for the hachures are then drawn in pencil, as shown in Fig. 3. These guide lines should always cross the contours at 90 deg. They will, for the most part, be slightly curved lines. They should be relatively close together, where contours are irregular and crooked, and can be spaced much farther apart where contours are nearly straight.

Since the contour interval divided by the scaled-map distance between adjacent contours gives the tangent of the angle of slope, the draftsman, by referring to the shading scaled as shown in Fig. 18, can determine the weight and spacing of hachures for any given portion of the map. Changes in spacing and weight must always be gradual to avoid a patchwork effect in shading. In hachuring a slope, portions having the same shading should be completely inked, working from the top contour down, as shown in Fig. 4, Plate 6. They should never be inked as shown in Fig. 5 of the same plate.

Correct hill shading is very difficult for the beginner, and he should observe the following precautions until he has developed some skill in this line:

1. Avoid long vertical strokes. Use auxilliary contours to cut down the contour interval.

2. Use plenty of guide lines.

3. Where contours are sharply curved or crooked, make your hachures follow the curve of the guide lines. Straight hachures under these conditions will make your work seem stiff and mechanical.

4. To change weight of line, use heavier pen and do not try to get all the various degrees of shading with the same pen and varied pressure.

5. Keep pens clean and do not hurry, as this is almost sure to result in poor work.

Topographic Sketching in Oblique.—Many topographic features, *viz.*, rocky terrain, gravel pits, stone quarries, eroded bluffs, and stream banks, can be strikingly shown in oblique, provided the draftsman has a certain degree of skill. The above features, well sketched, add much to the appearance of the map. It is an art easily attained, provided the draftsman will persevere in sketching a few such features from nature until he has formed a mental picture of its representation. Plate 3 shows several examples of oblique sketching that are pictorial and need no explanation.

Miscellaneous Symbols.— Many branches of engineering, as railway, navigation, military, oil and gas industry, U. S. Geological Survey, etc., have adopted symbols of both culture and vegetation which are not in common use, but are used in these industries for deciphering maps. They are simple and mechanical in their execution, and no special instruction to the student is necessary.

The authors have grouped in Plates 7 to 19 symbols most commonly used in general topographic work.

Plates 20 to 26 show the symbols adopted by the American Railway Engineering Association and have been revised to date.

Plates 28 to 42 show military symbols and special symbols developed by various branches of the engineering profession and are self-explanatory.

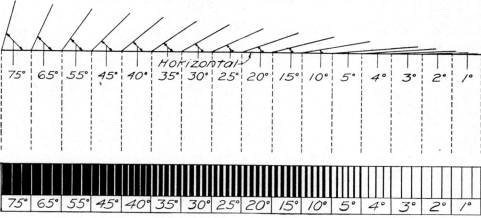

FIG. 18.

Plates 43 to 46 are suggested plates for student practice. Plate 45 is left partially blank as an exercise in toning the vegetation symbols to an equal shade with the hydrographic features.

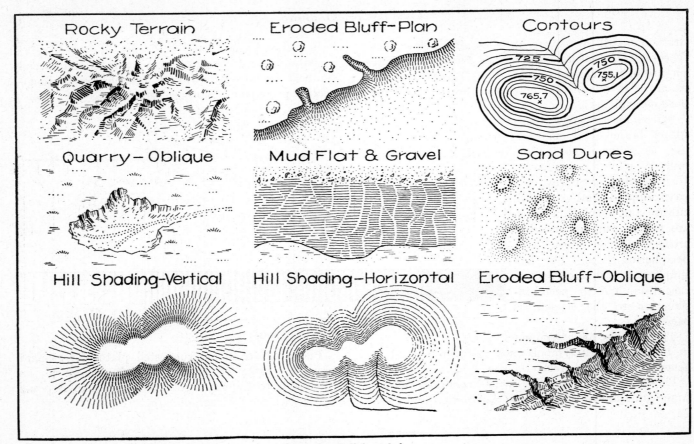

PLATE 3.—Relief symbols.

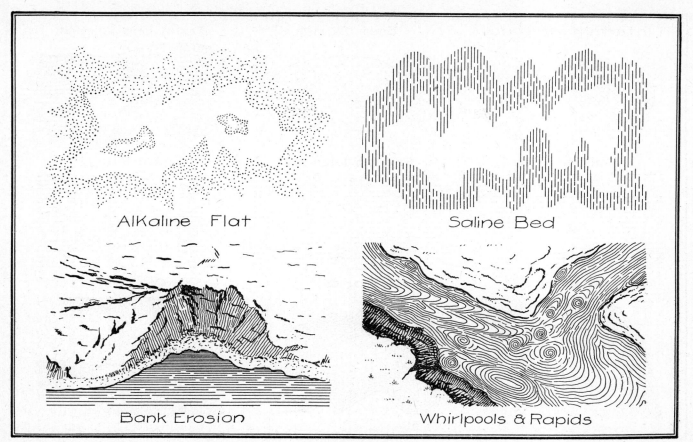

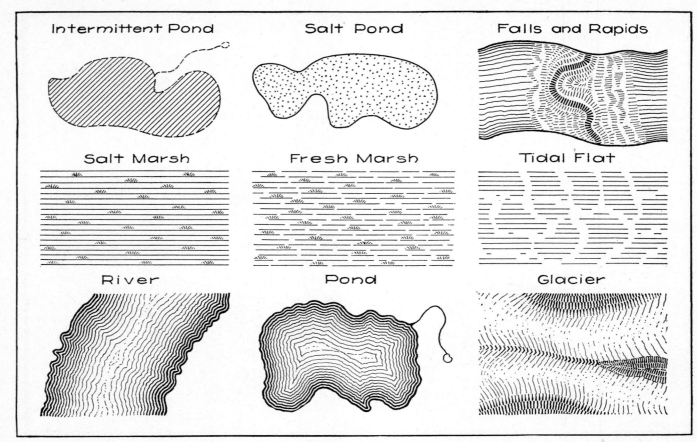

PLATE 5.—Hydrographic symbols.

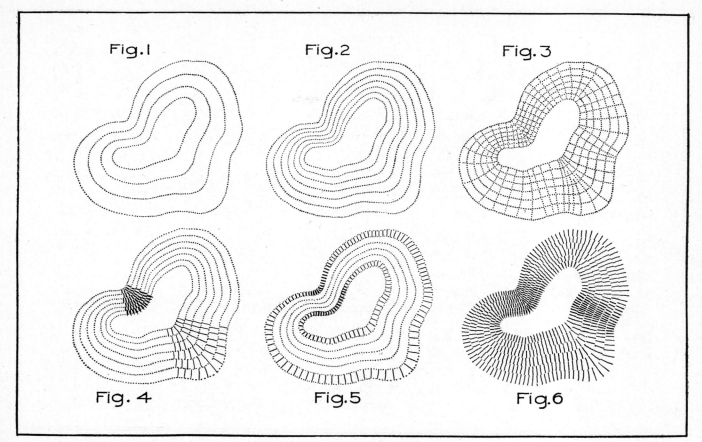

PLATE 6.—Development of hill shading.

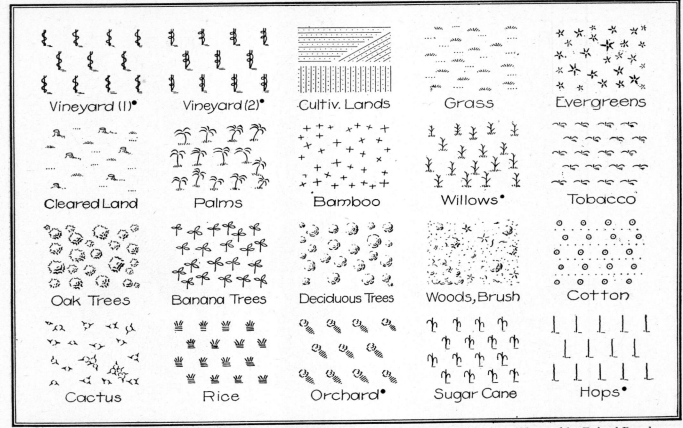

PLATE 7.—Vegetation symbols. (By permission of *Federal Board of Surveys and Maps, U. S. A.*). *Not used by Federal Board.

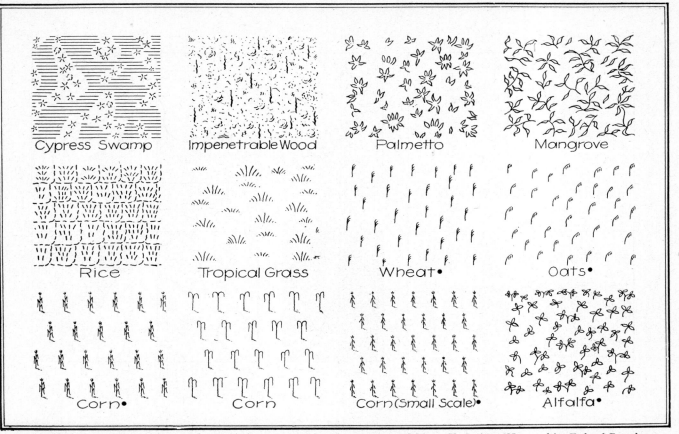

PLATE 8.—Vegetation symbols. (*By permission of Federal Board of Surveys and Maps, U. S. A.*) •Not used by Federal Board.

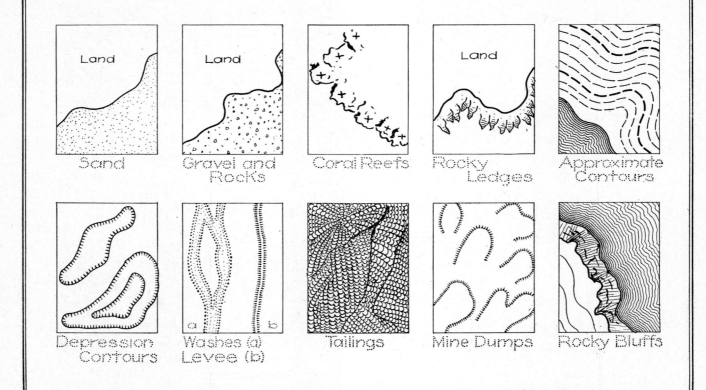

PLATE 9.—(*By permission of Federal Board of Surveys and Maps, U. S. A.*)

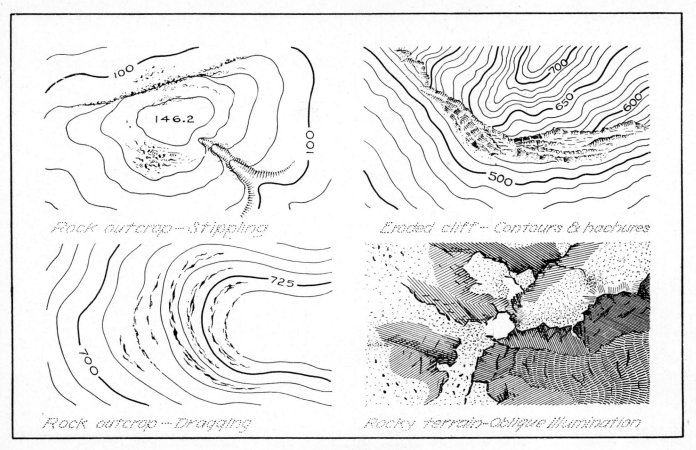

PLATE 10.—Rock drawing.

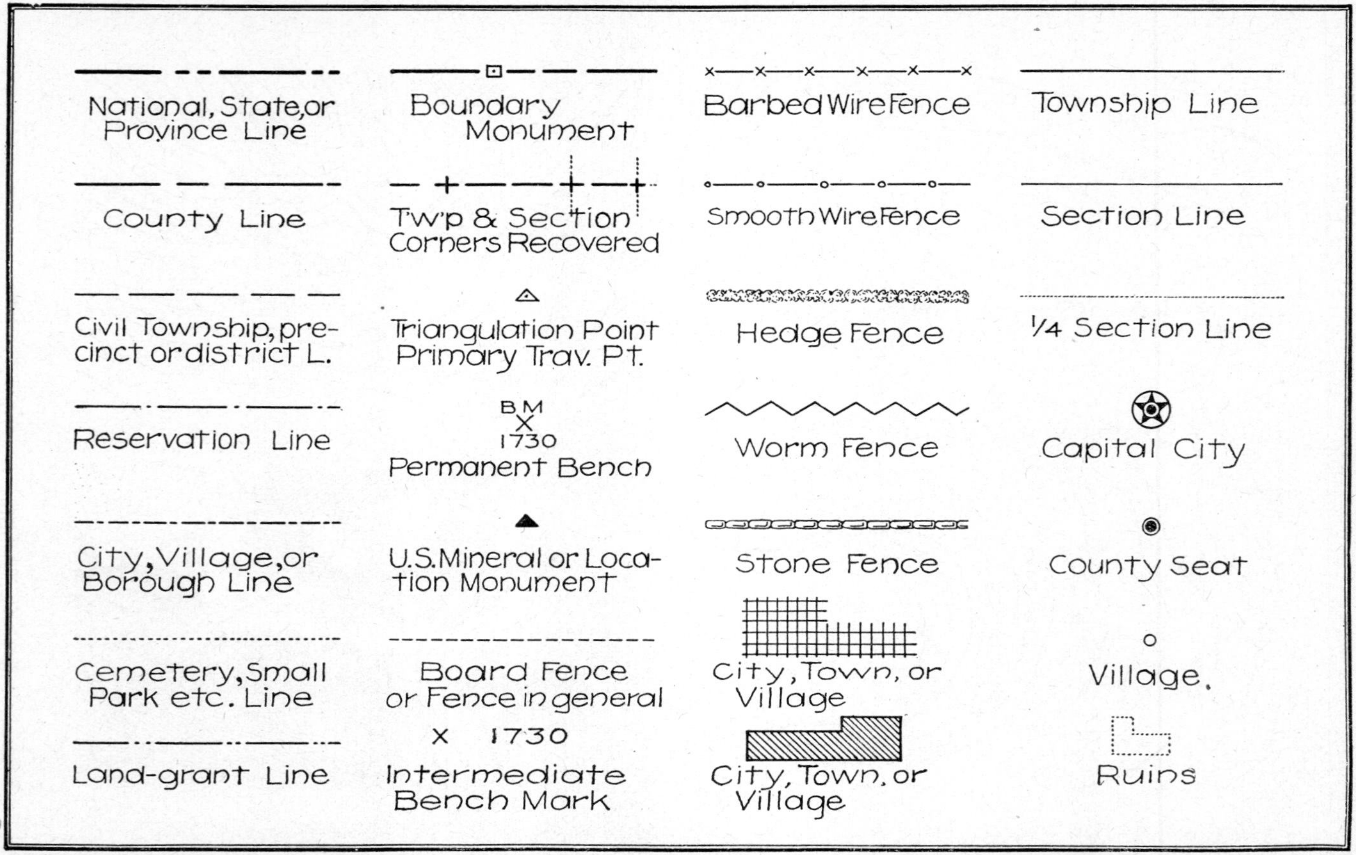

Symbol	Name	Symbol	Name	Symbol	Name	Symbol	Name
——— — —— —	National, State, or Province Line	— □ — —	Boundary Monument	×—×—×—×—×	Barbed Wire Fence	————————	Township Line
————————	County Line	—+—·—+—·—	Tw'p & Section Corners Recovered	—○——○——○—	Smooth Wire Fence	———————	Section Line
— — — — —	Civil Township, precinct or district L.	△ Triangulation Point Primary Trav. Pt.		≈≈≈≈≈≈≈ Hedge Fence		·············	¼ Section Line
—··—··—··—	Reservation Line	B M × 1730 Permanent Bench		∿∿∿∿∿ Worm Fence		✪	Capital City
—···—···—···—	City, Village, or Borough Line	▲ U.S. Mineral or Location Monument		◌◌◌◌◌◌ Stone Fence		⊙	County Seat
·············	Cemetery, Small Park etc. Line	———————— Board Fence or Fence in general		▦ City, Town, or Village		○	Village
—·············—·············—	Land-grant Line	× 1730 Intermediate Bench Mark		▨ City, Town, or Village		⌐⌐	Ruins

PLATE 11.—Boundary, fence and miscellaneous symbols. (*By permission of Federal Board of Surveys and Maps, U. S. A.*)

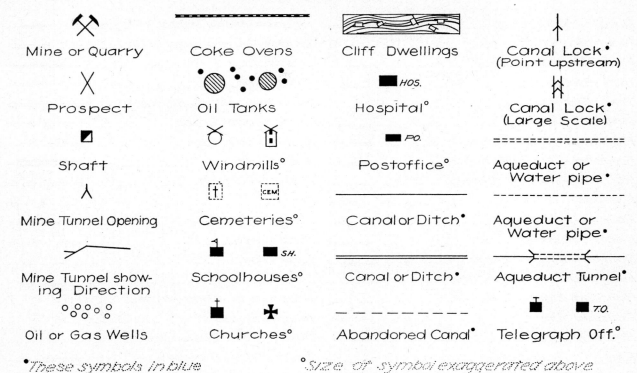

PLATE 12.—(By permission of Federal Board of Surveys and Maps, U. S. A.)

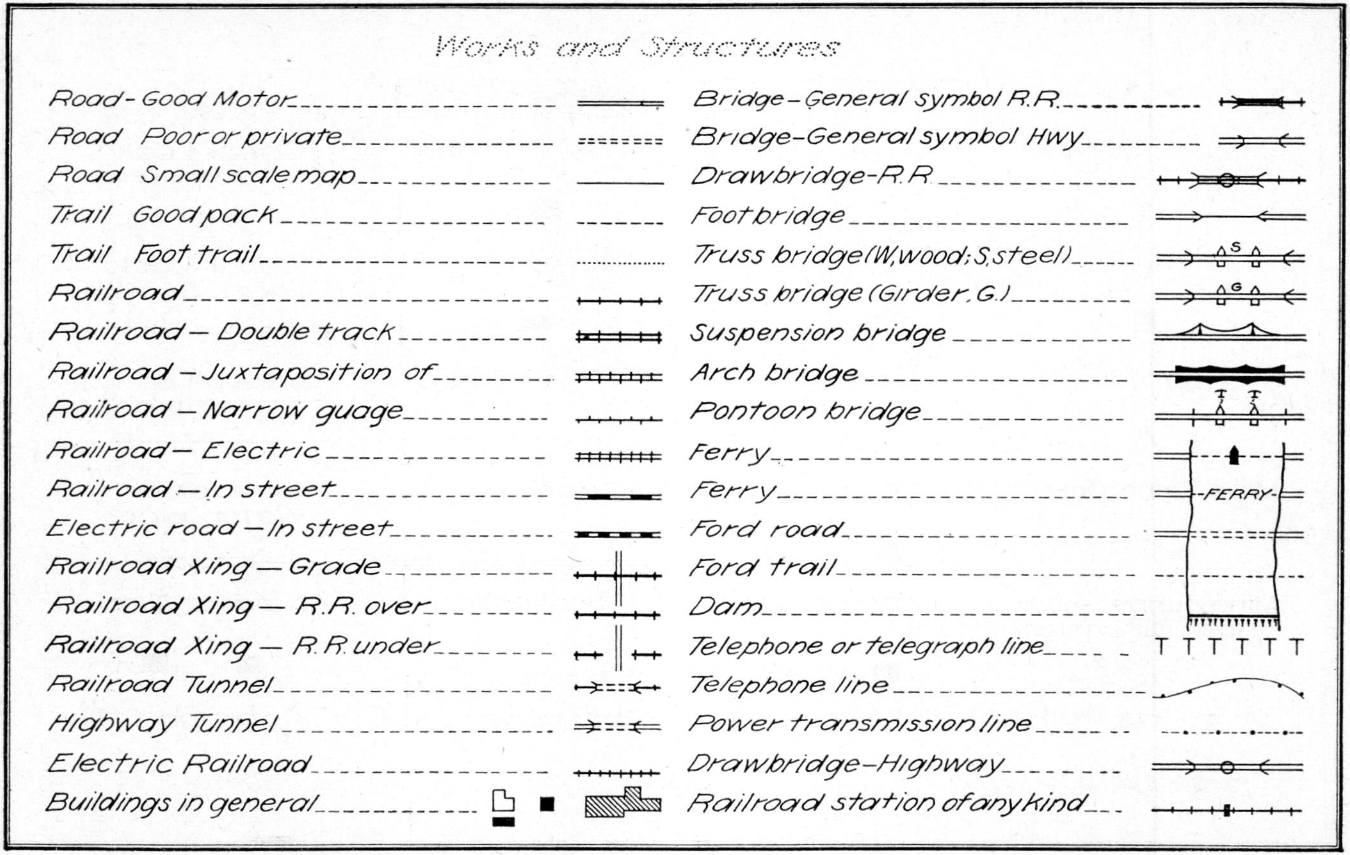

Works and Structures

Road – Good Motor	————		Bridge – General symbol R.R.	—▷▬◁—
Road Poor or private	========		Bridge – General symbol Hwy	—▷—◁—
Road Small scale map	————		Drawbridge – R.R.	—▷⊙◁—
Trail Good pack	‒ ‒ ‒ ‒		Footbridge	—▷ ◁—
Trail Foot trail	··········		Truss bridge (W, wood; S, steel)	—▷△ˢ△◁—
Railroad	‒†‒†‒†‒		Truss bridge (Girder, G.)	—▷△ᴳ△◁—
Railroad – Double track	⊢⊢⊢⊢		Suspension bridge	⌒⌒⌒
Railroad – Juxtaposition of	≢≢≢≢		Arch bridge	▬▬▬
Railroad – Narrow guage	‖‖‖‖		Pontoon bridge	—△△—
Railroad – Electric	╫╫╫╫		Ferry	– – ▲ – –
Railroad – In street	▬▬▬		Ferry	‒‒FERRY‒‒
Electric road – In street	▬▬▬		Ford road	– – – –
Railroad Xing – Grade	─╫─		Ford trail	– · – · –
Railroad Xing – R.R. over	─‖─		Dam	▬▬▬▬
Railroad Xing – R.R. under	←‖→		Telephone or telegraph line	T T T T T
Railroad Tunnel	⇉═⇇		Telephone line	⌒⌒⌒
Highway Tunnel	⇉═⇇		Power transmission line	– · – · –
Electric Railroad	╫╫╫╫		Drawbridge – Highway	—▷⊙◁—
Buildings in general	▯ ▪	▨	Railroad station of any kind	‒†‒▪‒†‒

PLATE 13.—(*By permission of Federal Board of Surveys and Maps U. S. A.*)

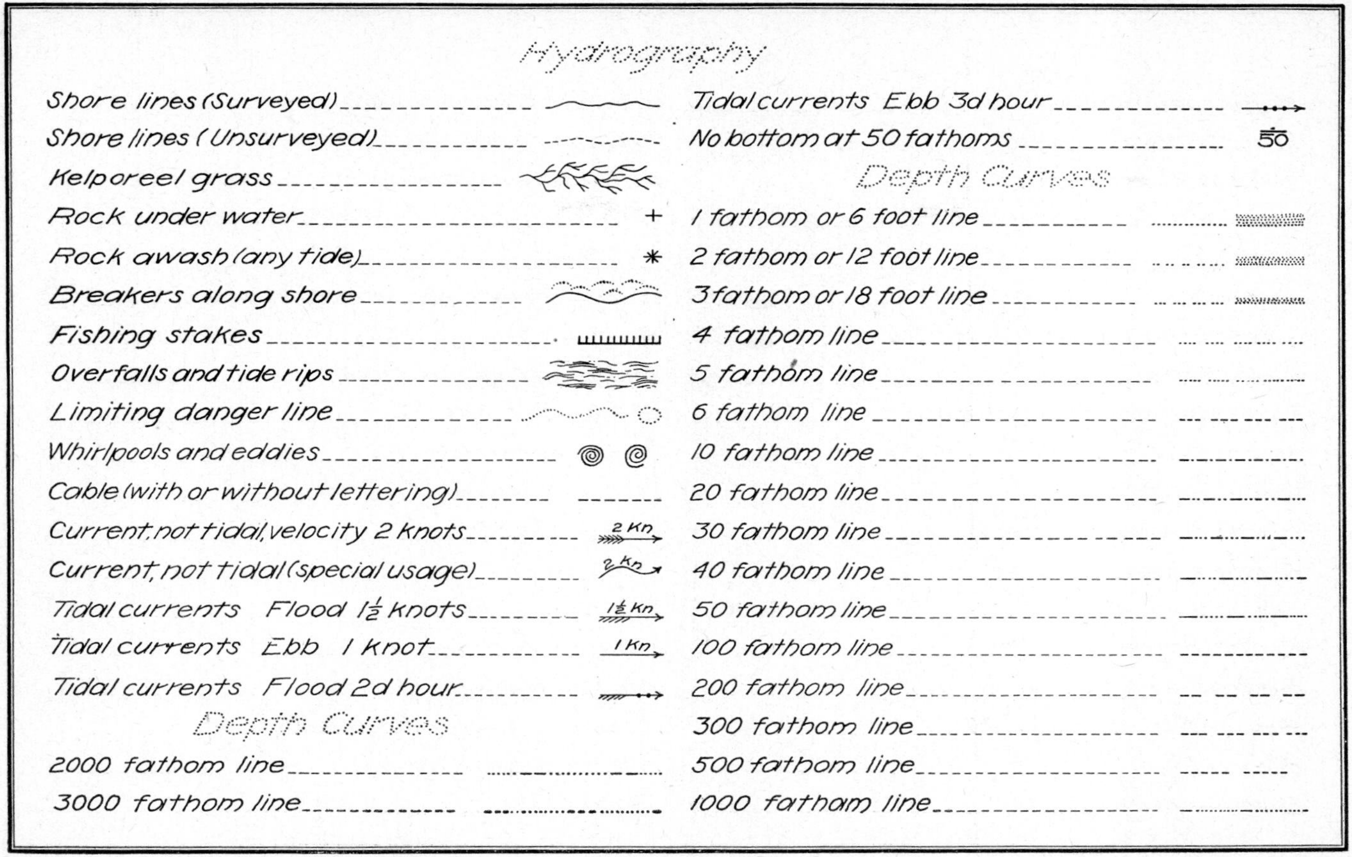

Hydrography

Shore lines (Surveyed)	Tidal currents Ebb 3d hour
Shore lines (Unsurveyed)	No bottom at 50 fathoms
Kelp or eel grass	*Depth Curves*
Rock under water	1 fathom or 6 foot line
Rock awash (any tide)	2 fathom or 12 foot line
Breakers along shore	3 fathom or 18 foot line
Fishing stakes	4 fathom line
Overfalls and tide rips	5 fathom line
Limiting danger line	6 fathom line
Whirlpools and eddies	10 fathom line
Cable (with or without lettering)	20 fathom line
Current, not tidal, velocity 2 knots	30 fathom line
Current, not tidal (special usage)	40 fathom line
Tidal currents Flood 1½ knots	50 fathom line
Tidal currents Ebb 1 knot	100 fathom line
Tidal currents Flood 2d hour	200 fathom line
Depth Curves	300 fathom line
2000 fathom line	500 fathom line
3000 fathom line	1000 fathom line

PLATE 14.—(*By permission of Federal Board of Surveys and Maps, U. S. A.*)

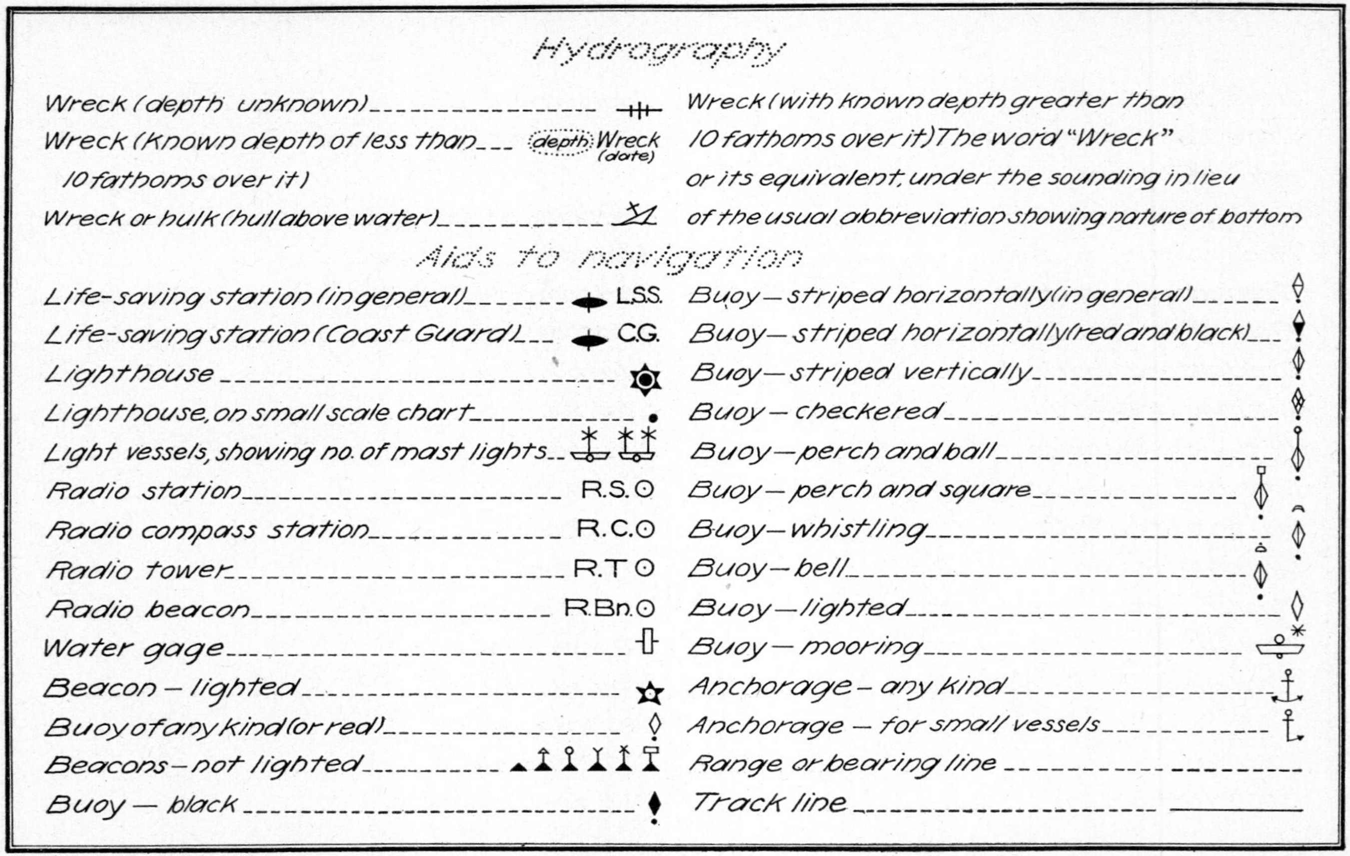

Hydrography

Wreck (depth unknown) ---------------- ⊬⊦

Wreck (known depth of less than 10 fathoms over it) ---- (depth) Wreck (date)

Wreck or hulk (hull above water) ---------- ⋈

Aids to navigation

Life-saving station (in general) ------ ◆ L.S.S.

Life-saving station (Coast Guard) --- ◆ C.G.

Lighthouse ---------------------- ✹

Lighthouse, on small scale chart ------ •

Light vessels, showing no. of mast lights -- ⚓ ⚓

Radio station ------------------ R.S. ⊙

Radio compass station ----------- R.C. ⊙

Radio tower -------------------- R.T ⊙

Radio beacon ------------------ R.Bn. ⊙

Water gage -------------------- ⌶

Beacon — lighted --------------- ☆

Buoy of any kind (or red) -------- ◇

Beacons — not lighted ------- ▲ ⏄ ⏄ ⏄ ⏄

Buoy — black ------------------ ♦

Wreck (with known depth greater than 10 fathoms over it) The word "Wreck" or its equivalent, under the sounding in lieu of the usual abbreviation showing nature of bottom

Buoy — striped horizontally (in general) ---- ◇

Buoy — striped horizontally (red and black) --- ◆

Buoy — striped vertically ---------- ◇

Buoy — checkered ---------------- ◇

Buoy — perch and ball ------------ ◇

Buoy — perch and square ---------- ◇

Buoy — whistling ---------------- ◇

Buoy — bell -------------------- ◇

Buoy — lighted ------------------ ◇

Buoy — mooring ---------------- ⛵

Anchorage — any kind ------------ ⚓

Anchorage — for small vessels ------ ⚓

Range or bearing line ------------

Track line ----------------------

PLATE 15.—(By permission of Federal Board of Surveys and Maps, U. S. A.)

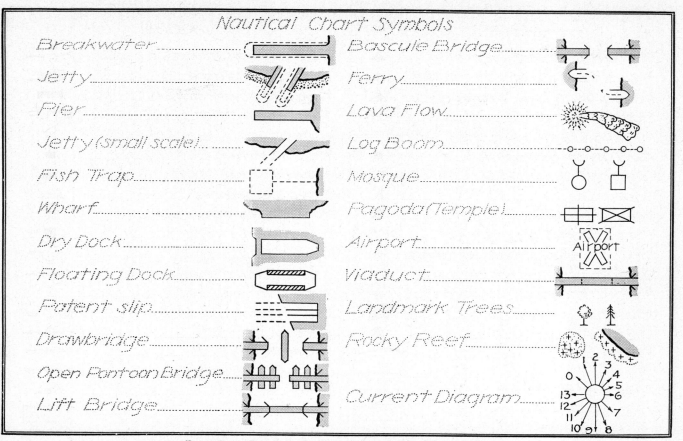

PLATE 16.—(*By permission of U. S. Coast and Geodetic Survey.*)

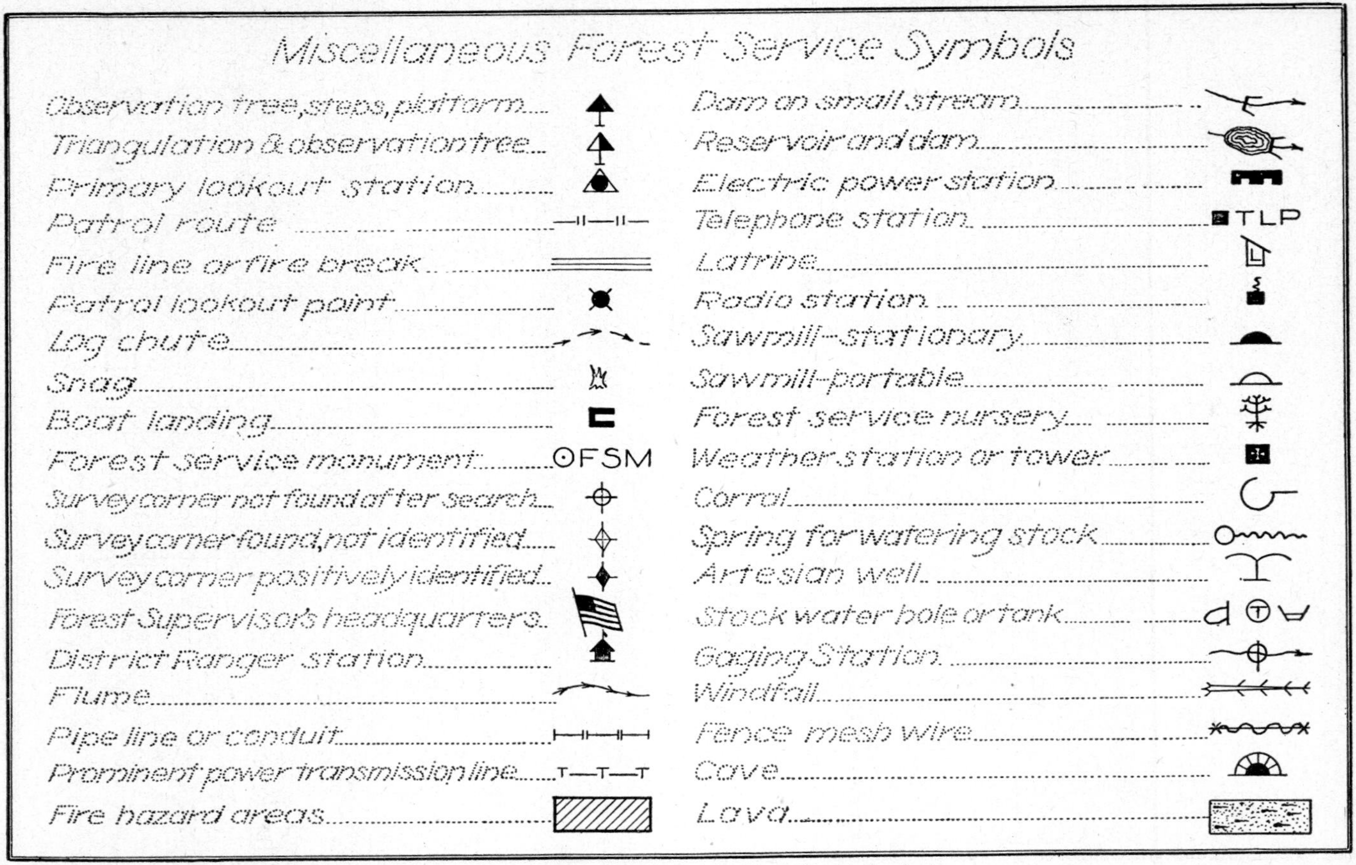

Miscellaneous Forest Service Symbols

Symbol		Symbol	
Observation tree, steps, platform	▲	Dam on small stream	
Triangulation & observation tree	▲	Reservoir and dam	
Primary lookout station	▲	Electric power station	
Patrol route	—‖—‖—	Telephone station	■ TLP
Fire line or fire break	≡	Latrine	
Patrol lookout point	✳	Radio station	
Log chute	→→	Sawmill—stationary	
Snag		Sawmill—portable	
Boat landing	⊏	Forest service nursery	
Forest service monument	⊙ FSM	Weather station or tower	
Survey corner not found after search	✧	Corral	
Survey corner found, not identified	✦	Spring for watering stock	
Survey corner positively identified	✦	Artesian well	
Forest Supervisor's headquarters	⚑	Stock water hole or tank	
District Ranger station	⌂	Gaging Station	
Flume		Windfall	
Pipe line or conduit	⊢⊢⊢	Fence mesh wire	
Prominent power transmission line	T—T—T	Cave	
Fire hazard areas	▨	Lava	

PLATE 17.—(*By permission of the U. S. Forest Service.*)

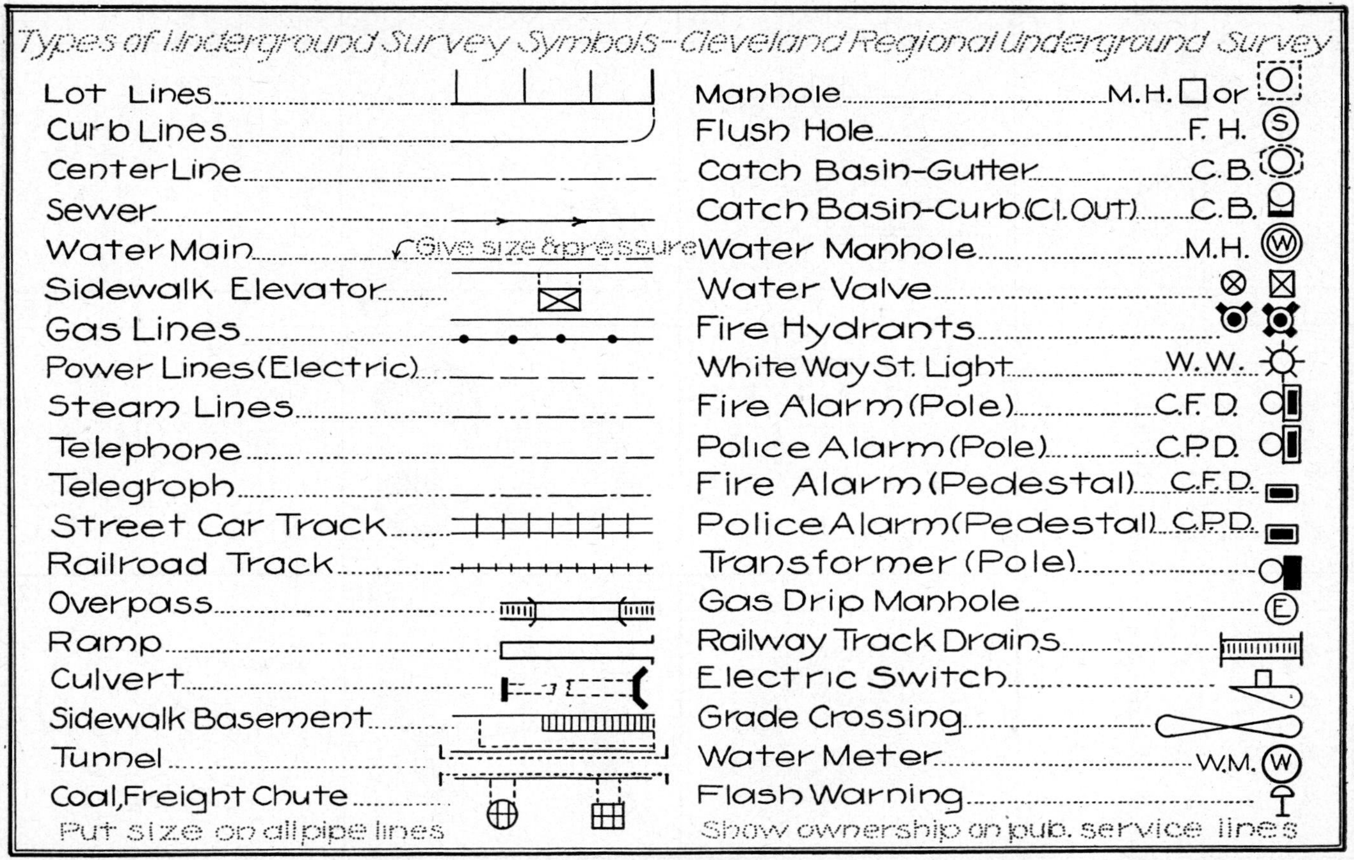

Types of Underground Survey Symbols — Cleveland Regional Underground Survey

Feature	Symbol	Feature	Symbol
Lot Lines		Manhole	M.H. □ or ⬚
Curb Lines		Flush Hole	F. H. Ⓢ
Center Line		Catch Basin-Gutter	C.B. ⊚
Sewer		Catch Basin-Curb (Cl. Out)	C.B.
Water Main	Give size & pressure	Water Manhole	M.H. Ⓦ
Sidewalk Elevator	⊠	Water Valve	⊗ ⊠
Gas Lines	•—•—•—	Fire Hydrants	
Power Lines (Electric)		White Way St. Light	W.W. ☼
Steam Lines		Fire Alarm (Pole)	C.F.D.
Telephone		Police Alarm (Pole)	C.P.D.
Telegraph		Fire Alarm (Pedestal)	C.F.D. ▬
Street Car Track	╫╫╫╫╫	Police Alarm (Pedestal)	C.P.D. ▬
Railroad Track	++++++	Transformer (Pole)	
Overpass		Gas Drip Manhole	Ⓔ
Ramp		Railway Track Drains	
Culvert		Electric Switch	
Sidewalk Basement		Grade Crossing	
Tunnel		Water Meter	W.M. Ⓦ
Coal, Freight Chute		Flash Warning	
Put size on all pipe lines		Show ownership on pub. service lines	

PLATE 18.—(*By permission of Cleveland Regional Underground Survey.*)

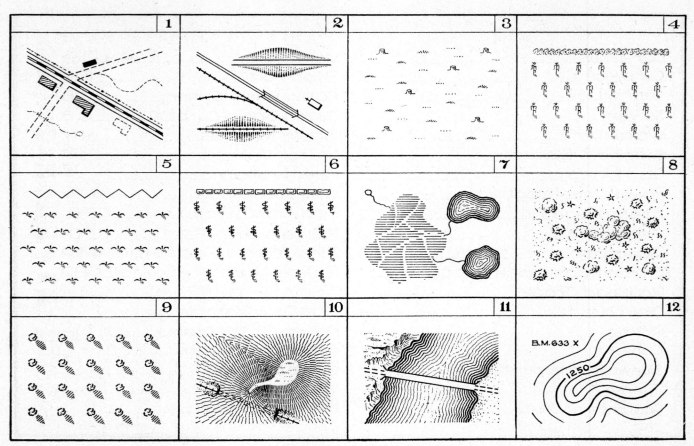

PLATE 19.—Topographic symbols.

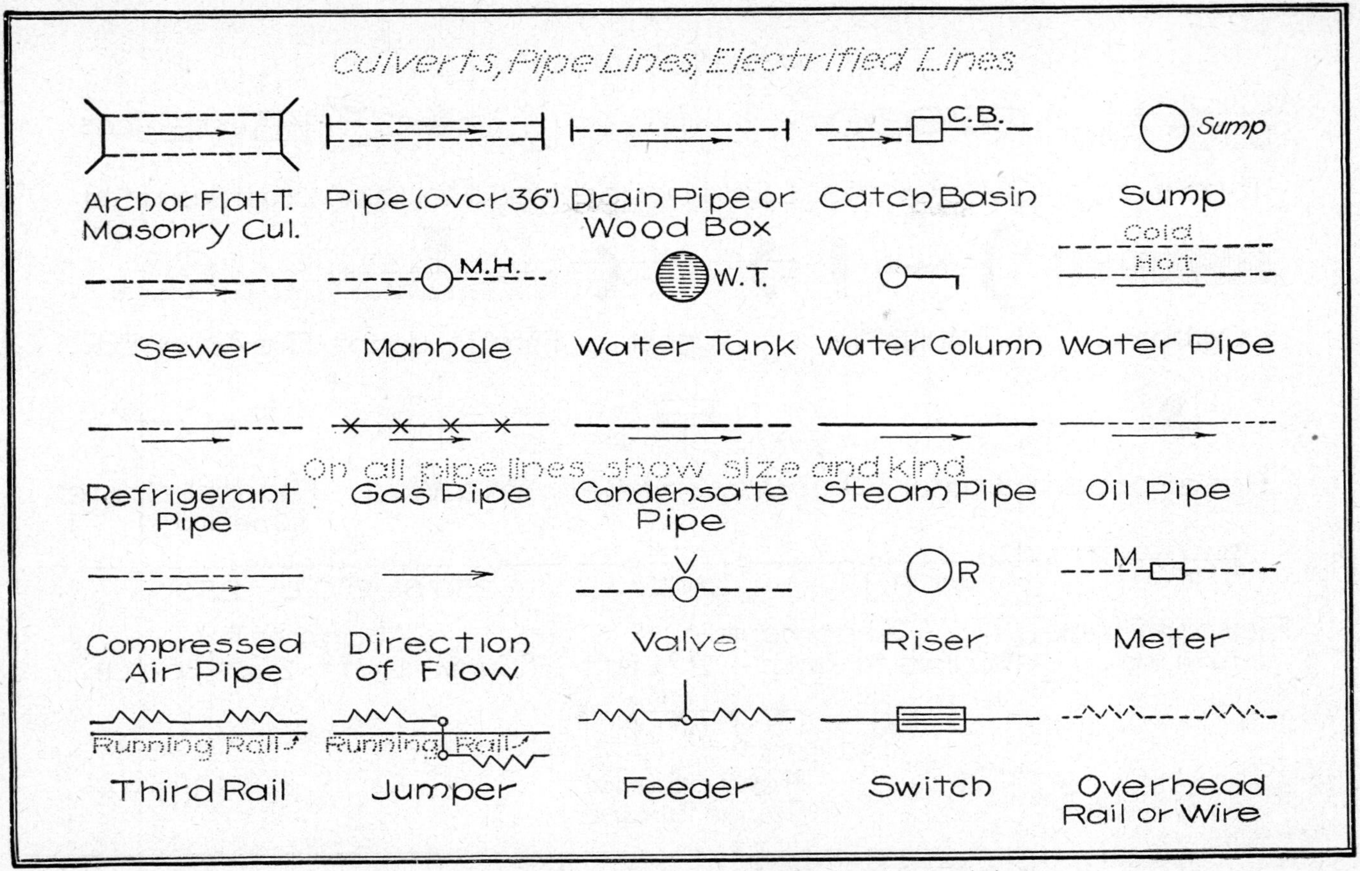

Culverts, Pipe Lines, Electrified Lines

Symbol	Symbol	Symbol	Symbol	Symbol
Arch or Flat T. Masonry Cul.	Pipe (over 36")	Drain Pipe or Wood Box	Catch Basin	Sump
Sewer	Manhole	Water Tank	Water Column	Water Pipe (Cold / Hot)
Refrigerant Pipe	Gas Pipe	Condensate Pipe	Steam Pipe	Oil Pipe
Compressed Air Pipe	Direction of Flow	Valve	Riser	Meter
Third Rail	Jumper	Feeder	Switch	Overhead Rail or Wire

On all pipe lines show size and kind

PLATE 20.—(*By permission of American Railway Engineering Association.*)

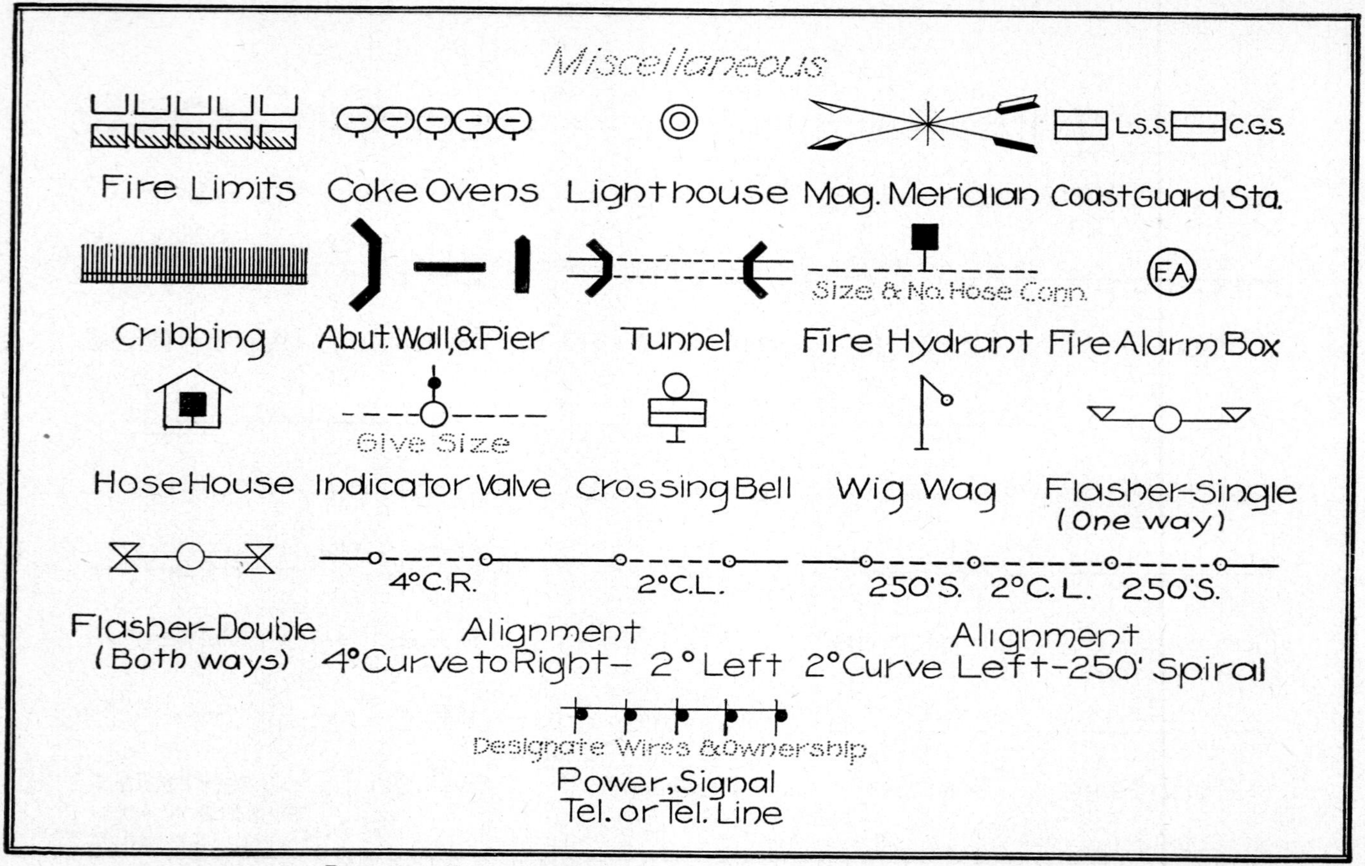

Miscellaneous

Fire Limits	Coke Ovens	Lighthouse	Mag. Meridian	Coast Guard Sta. (L.S.S. / C.G.S.)
Cribbing	Abut. Wall & Pier	Tunnel	Fire Hydrant (Size & No. Hose Conn.)	Fire Alarm Box (F.A.)
Hose House	Indicator Valve (Give Size)	Crossing Bell	Wig Wag	Flasher–Single (One way)
Flasher–Double (Both ways)	Alignment 4° Curve to Right – 2° Left (4°C.R. / 2°C.L.)		Alignment 2° Curve Left – 250' Spiral (250'S. / 2°C.L. / 250'S.)	

Designate Wires & Ownership
Power, Signal
Tel. or Tel. Line

PLATE 21.—(By permission of American Railway Engineering Association.)

Fences, Highways, Crossings and Mines

Woven Wire Fence	Farm Gate	Board Fence	Barb Wire Fence	Worm Fence
Snow Fence	Hedge	Unfenced Property Line	Intertrack Fence	Public, Main Road
Private and Secondary R'ds	Trails	Street & Public Road Crossings	Private Road Crossing	Grade Crossing
Undergrade Crossing	Overhead Crossing	Crossing Gate	Turnstile	Cattle Guard
Operating Mine	Tunnel	Test Opening	Shaft	Coal Outcrop

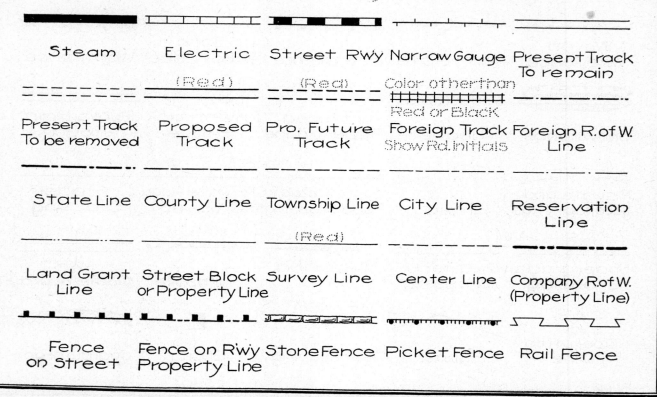

PLATE 23.—(*By permission of American Railway Engineering Association.*)

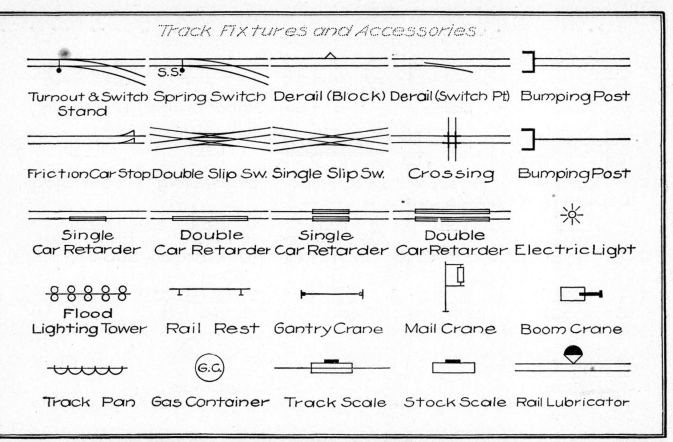

PLATE 24.—(*By permission of American Railway Engineering Association.*)

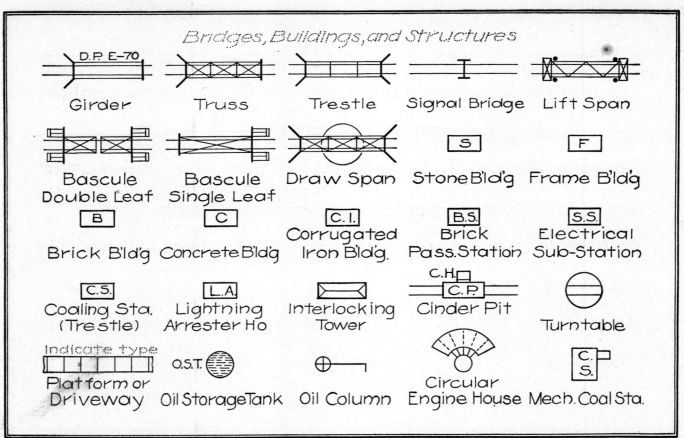

PLATE 25.—(By permission of American Railway Engineering Association.)

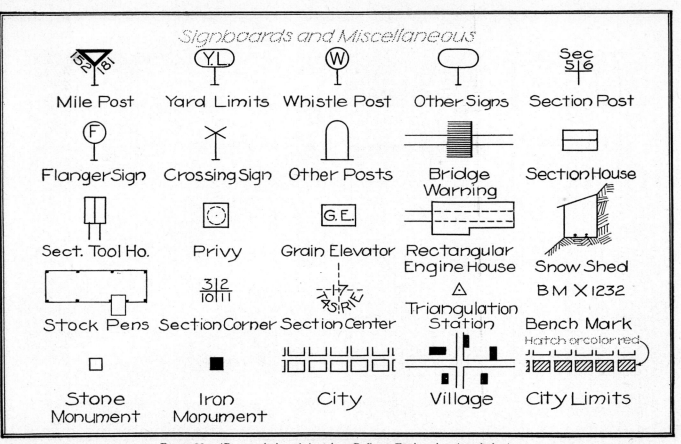

PLATE 26.—(*By permission of American Railway Engineering Association.*)

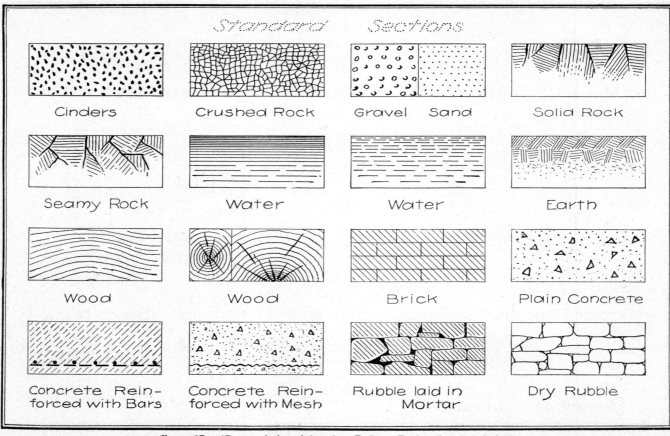

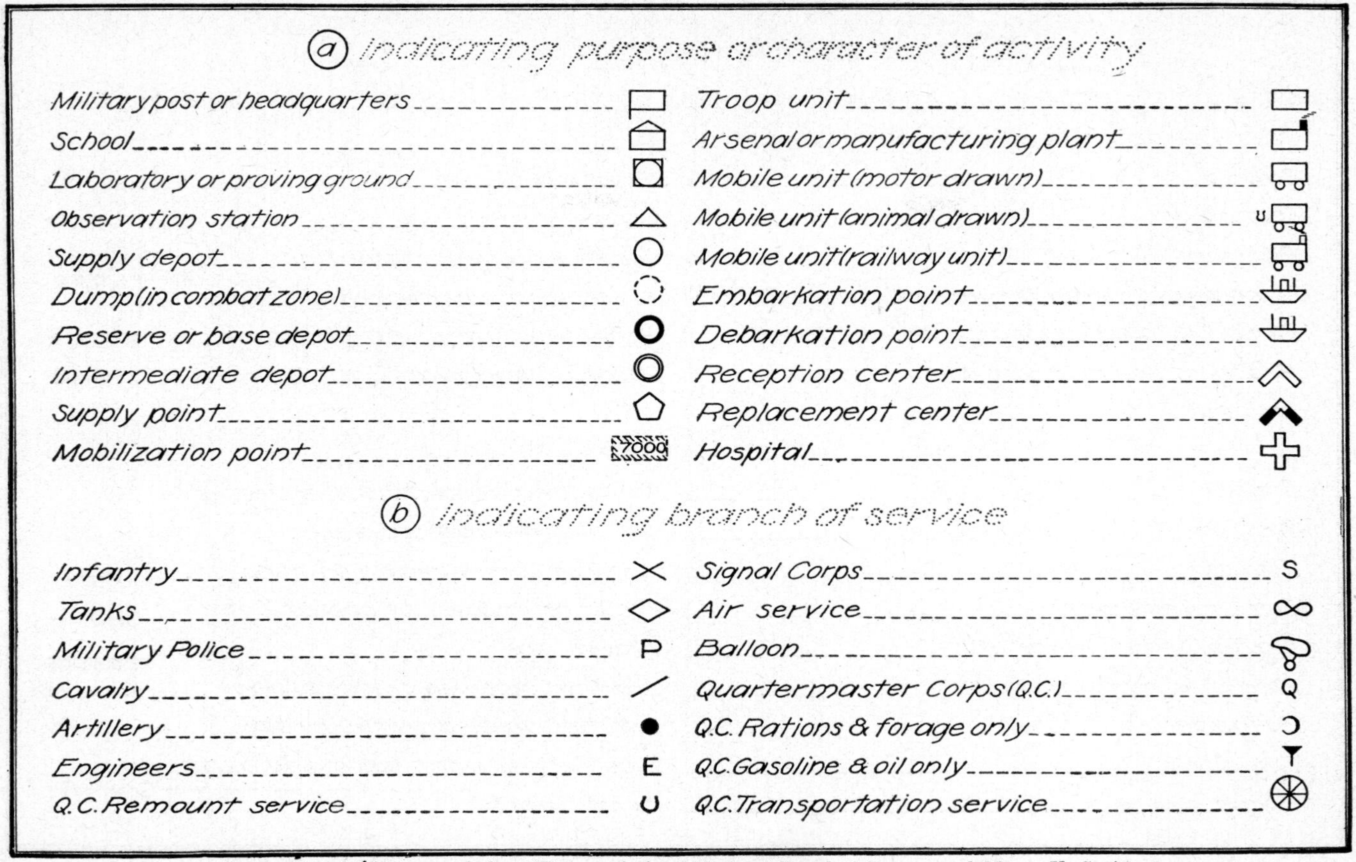

(a) Indicating purpose or character of activity

Military post or headquarters	▢	Troop unit	▭
School	⌂	Arsenal or manufacturing plant	(symbol)
Laboratory or proving ground	◯ (oval)	Mobile unit (motor drawn)	(symbol)
Observation station	△	Mobile unit (animal drawn)	(symbol)
Supply depot	○	Mobile unit (railway unit)	(symbol)
Dump (in combat zone)	◌	Embarkation point	(ship symbol)
Reserve or base depot	●	Debarkation point	(ship symbol)
Intermediate depot	◎	Reception center	⋀
Supply point	⬠	Replacement center	▲
Mobilization point	[7000]	Hospital	✚

(b) Indicating branch of service

Infantry	×	Signal Corps	S
Tanks	◇	Air service	∞
Military Police	P	Balloon	(symbol)
Cavalry	/	Quartermaster Corps (Q.C.)	Q
Artillery	●	Q.C. Rations & forage only	C
Engineers	E	Q.C. Gasoline & oil only	⊤
Q.C. Remount service	U	Q.C. Transportation service	⊛

PLATE 28.—Military symbols. (By permission of Federal Board of Surveys and Maps, U. S. A.)

(c) Indicating size of units

Note: These symbols will be placed above the symbols shown in "a" or used for indicating boundaries as shown in "d" below

Squad	•	Division	xx
Section	••	Corps	xxx
Platoon	•••	Corps area, department, or section communications zone	ooo
Company, troop, battery, or Air Service squad'n	I		
Battalion, Cavalry squad'n, or Air Serv. group	II	Army or communications zone	xxxx
Regiment or Air Service wing	III	General headquarters	GHQ
Brigade	x		

(d) Boundaries

Squad	——·——	Corps area, department, or section communications zone	——ooo——
Section	——··——		
Platoon	——···——	Army or communications zone	——xxxx——
Company or similar unit	—I——I—	Rear of theater of operations	——GHQ——
Battalion or similar unit	—II——II—	Front line	▬▬▬▬▬
Regiment or similar unit	—III——III—	Limit of wheeled traffic by night	—NT——NT—
Brigade	—x——x—	Limit of wheeled traffic by day	—DY——DY—
Division	—xx——xx—	Line beyond which gas masks must be at "alert"	—G——G—
Corps	—xxx——xxx—		

PLATE 29.—Military symbols. (*By permission of Federal Board of Surveys and Maps, U. S. A.*)

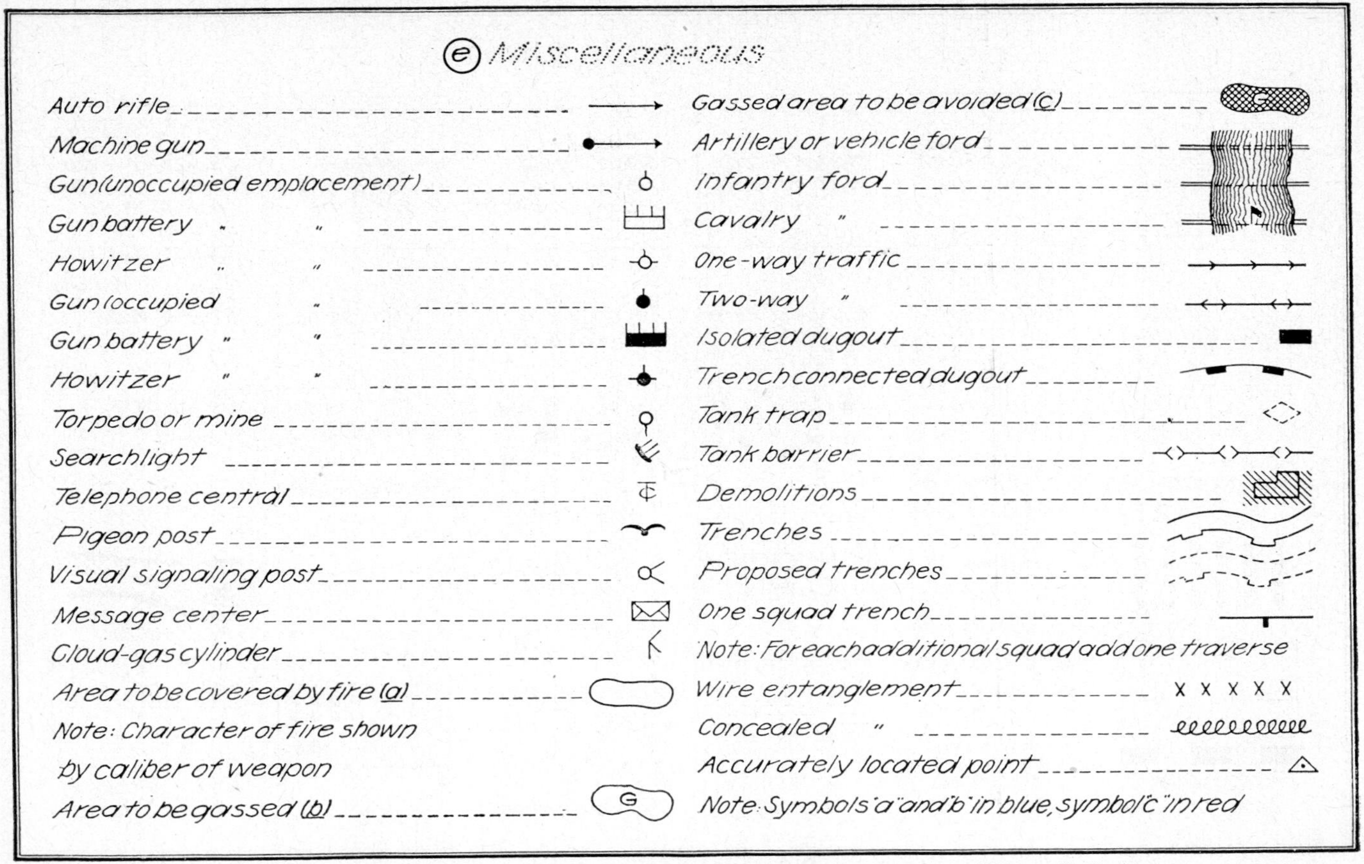

(e) Miscellaneous

Auto rifle	→	Gassed area to be avoided (c)	▨
Machine gun	•→	Artillery or vehicle ford	≋
Gun (unoccupied emplacement)	○	Infantry ford	≋
Gun battery " "	⊔	Cavalry "	≋
Howitzer " "	⊖	One-way traffic	→→
Gun (occupied) " "	●	Two-way "	↔↔
Gun battery " "	⊔	Isolated dugout	▬
Howitzer " "	⊖	Trench connected dugout	▬⌒▬
Torpedo or mine	⚲	Tank trap	◇
Searchlight	⚔	Tank barrier	◇◇◇
Telephone central	⏀	Demolitions	▨
Pigeon post	⌒	Trenches	∼∼
Visual signaling post	⚔	Proposed trenches	⋯⋯
Message center	✉	One squad trench	—┼—
Cloud-gas cylinder	⎞	Note: For each additional squad add one traverse	
Area to be covered by fire (a)	⬭	Wire entanglement	× × × × ×
Note: Character of fire shown by caliber of weapon		Concealed "	ℓℓℓℓℓℓℓℓ
		Accurately located point	△
Area to be gassed (b)	Ⓖ	Note: Symbols "a" and "b" in blue, symbol "c" in red	

PLATE 30.—Military symbols. (*By permission of Federal Board of Surveys and Maps, U. S. A.*)

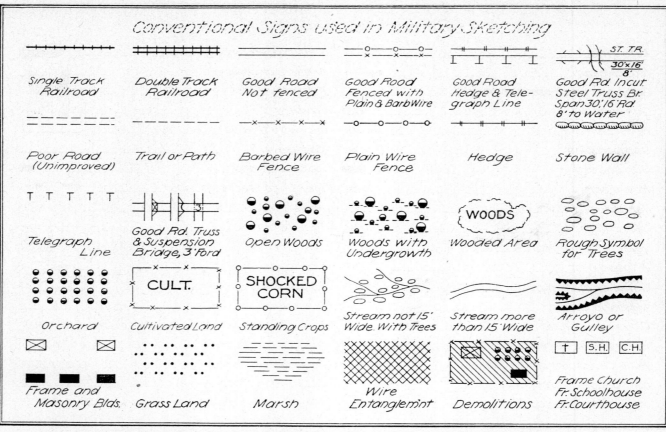

PLATE 31.—Symbols for military field sketching. (*By permission of War Department, U. S. A.*)

Unit	Symbol	Unit	Symbol	Unit	Symbol
Company C, 20th Infantry	C ⊠ 20	Battery F, 2nd Artillery	F ▪ 2	4th Signal Troop	S 4
3d Battalion, 8th Infantry	3 ⊠ 8	1st Battalion, 8th Artillery	1 ▪ 8	4th Wing Air Service	∞ 4
5th Infantry	⊠ 5	Antitank gun	◇	Balloon Company No. 5	⏂ 5
16th Infantry Brigade	⊠ 16	13th Artillery	▪ 13	Aid station Medical Dept.	+
8th Division (Infantry)	⊠ 8	1st Artillery Brigade	▪ 1	Battalion aid station	ǂ
Light Mortar (Infantry)	•− 1m	III Corps Area F.A. School	f 🏛 a	Regimental aid station	ǂ
Troop F, 2d Cavalry	F ⊟ 2	Engineer supply depot	Ⓔ	Sanitary Battalion 2d Med. Regt.	⊟ 2
1st Squadron, 4th Cavalry	1 ⊟ 4	Company A, 2nd Engineers	A Ⓔ 2	Sanitary Comp'y No. 44	⊟ 44
18th Cavalry	⊟ 18	Engineer dump	(Ⓔ)	Hospital Comp'y No. 2	⊟ 2h
1st Cavalry Brigade	⊟ 1	16th Signal Company	S 16	6th Gas Comp'y Chem. Warfare Ser	G 6
4th Cavalry Division	⊟ 4	17th Signal Battalion	S 17	Chemical Warfare Service Dump	(G)

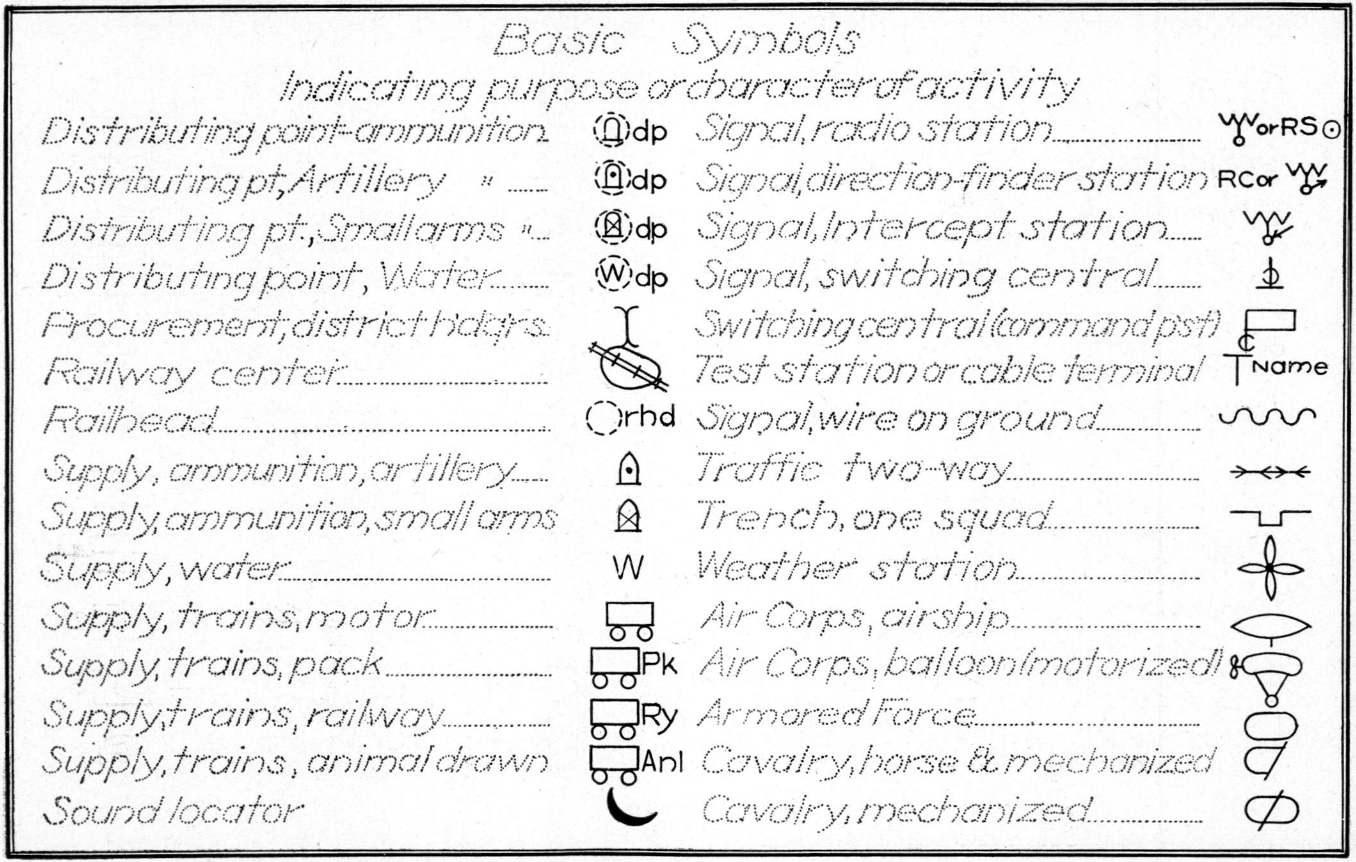

Basic Symbols
Indicating purpose or character of activity

Distributing point-ammunition	(A)dp	Signal, radio station	⋎⋎ or RS
Distributing pt, Artillery "	(A)dp	Signal, direction-finder station	RC or ⋎⋎
Distributing pt, Small arms "	(A)dp	Signal, intercept station	⋎⋎
Distributing point, Water	(W)dp	Signal, switching central	
Procurement, district hdqtrs.		Switching central (command post)	
Railway center		Test station or cable terminal	Name
Railhead	○rhd	Signal, wire on ground	
Supply, ammunition, artillery		Traffic two-way	
Supply, ammunition, small arms		Trench, one squad	
Supply, water	W	Weather station	
Supply, trains, motor		Air Corps, airship	
Supply, trains, pack	Pk	Air Corps, balloon (motorized)	
Supply, trains, railway	Ry	Armored Force	
Supply, trains, animal drawn	Anl	Cavalry, horse & mechanized	
Sound locator		Cavalry, mechanized	

PLATE 33.—(*By permission of War Department, U. S. A.*)

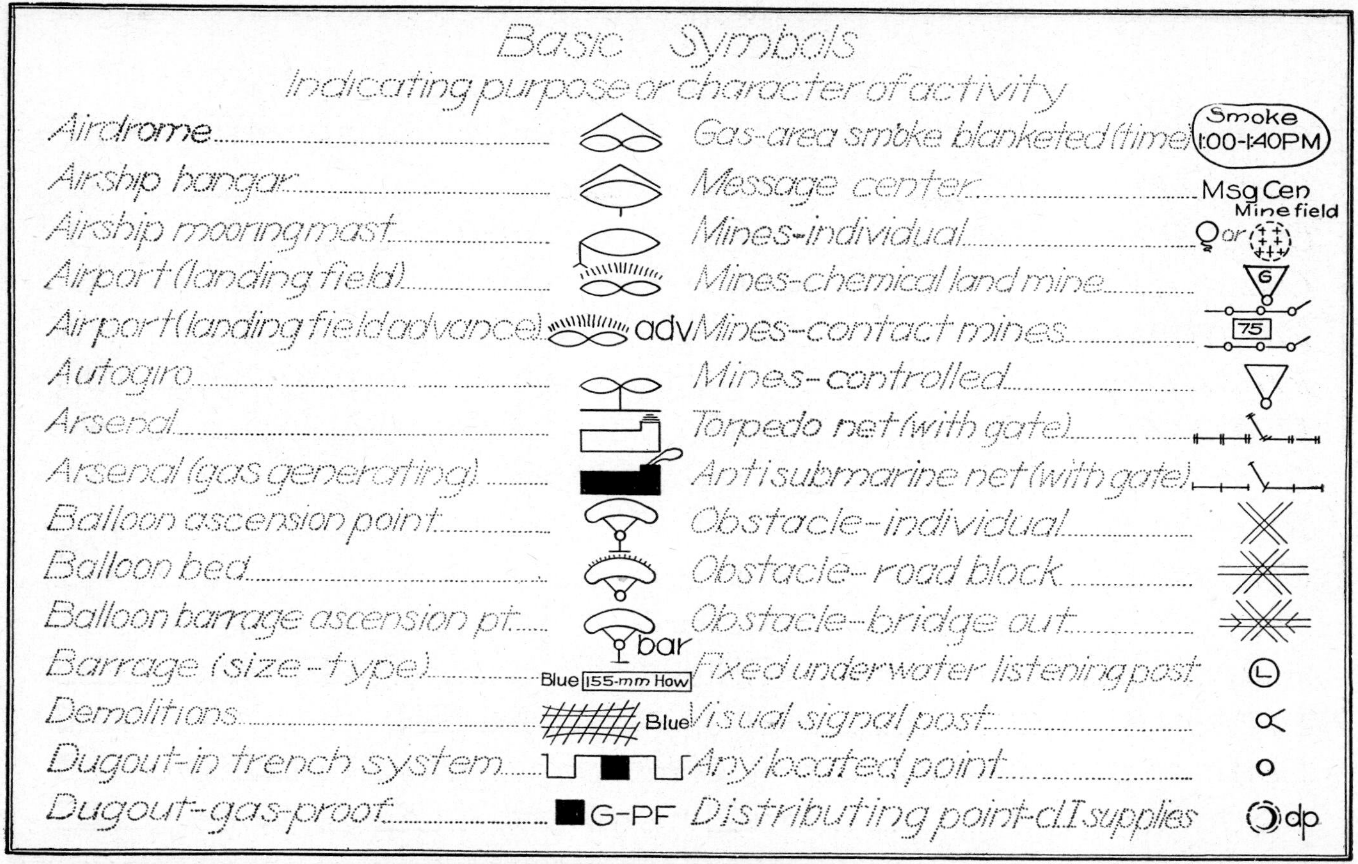
Basic Symbols
Indicating purpose or character of activity
Airdrome
Airship hangar
Airship mooring mast
Airport (landing field)
Airport (landing field advance)
adv
Autogiro
Arsenal
Arsenal (gas generating)
Balloon ascension point
Balloon bed
Balloon barrage ascension pt.
bar
Barrage (size-type)
Blue
155-mm How
Demolitions
Blue
Dugout-in trench system
Dugout-gas-proof
G-PF
Gas-area smoke blanketed (time)
Smoke 1:00-1:40PM
Message center
Msg Cen
Mines-individual
Mine field
or
Mines-chemical land mine
G
Mines-contact mines
75
Mines-controlled
Torpedo net (with gate)
Antisubmarine net (with gate)
Obstacle-individual
Obstacle-road block
Obstacle-bridge out
Fixed underwater listening post
L
Visual signal post
Any located point
Distributing point-cl. I supplies
dp

PLATE 34.—(*By permission of War Department, U. S. A.*)

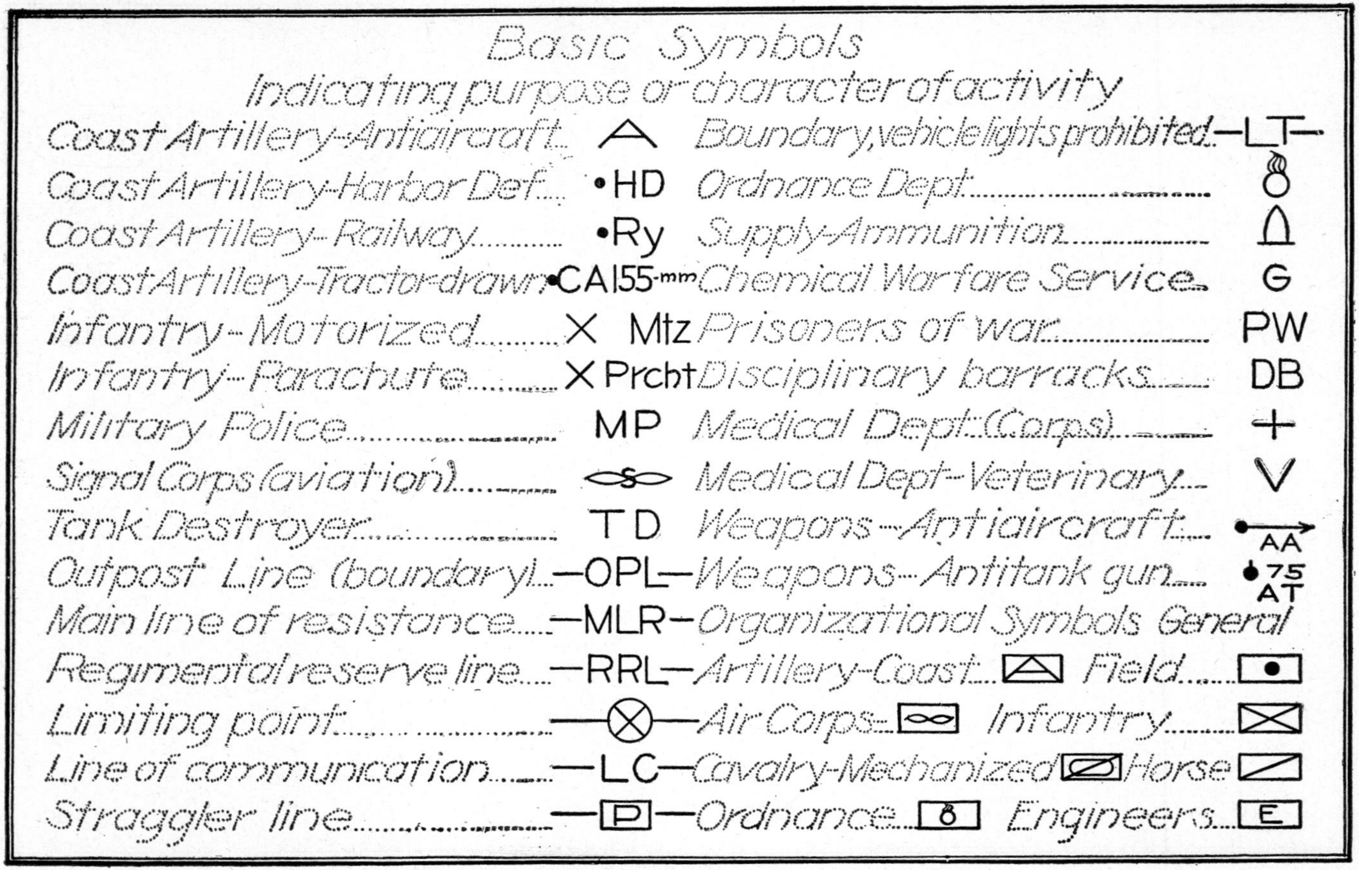

Basic Symbols
Indicating purpose or character of activity

Coast Artillery—Antiaircraft	△	Boundary, vehicle lights prohibited	—LT—
Coast Artillery—Harbor Def.	•HD	Ordnance Dept.	🔥
Coast Artillery—Railway	•Ry	Supply—Ammunition	△
Coast Artillery—Tractor-drawn	•CA 155-mm	Chemical Warfare Service	G
Infantry—Motorized	X Mtz	Prisoners of war	PW
Infantry—Parachute	X Prcht	Disciplinary barracks	DB
Military Police	MP	Medical Dept. (Corps)	+
Signal Corps (aviation)	⌒S⌒	Medical Dept—Veterinary	V
Tank Destroyer	TD	Weapons—Antiaircraft	•→ AA
Outpost Line (boundary)	—OPL—	Weapons—Antitank gun	•75 AT
Main line of resistance	—MLR—	Organizational Symbols General	
Regimental reserve line	—RRL—	Artillery—Coast ⌧ Field	•
Limiting point	—⊗—	Air Corps ∞ Infantry	⊠
Line of communication	—LC—	Cavalry—Mechanized ⌧ Horse	⧄
Straggler line	—P—	Ordnance 🔥 Engineers	E

PLATE 35.—(*By permission of War Department, U. S. A.*)

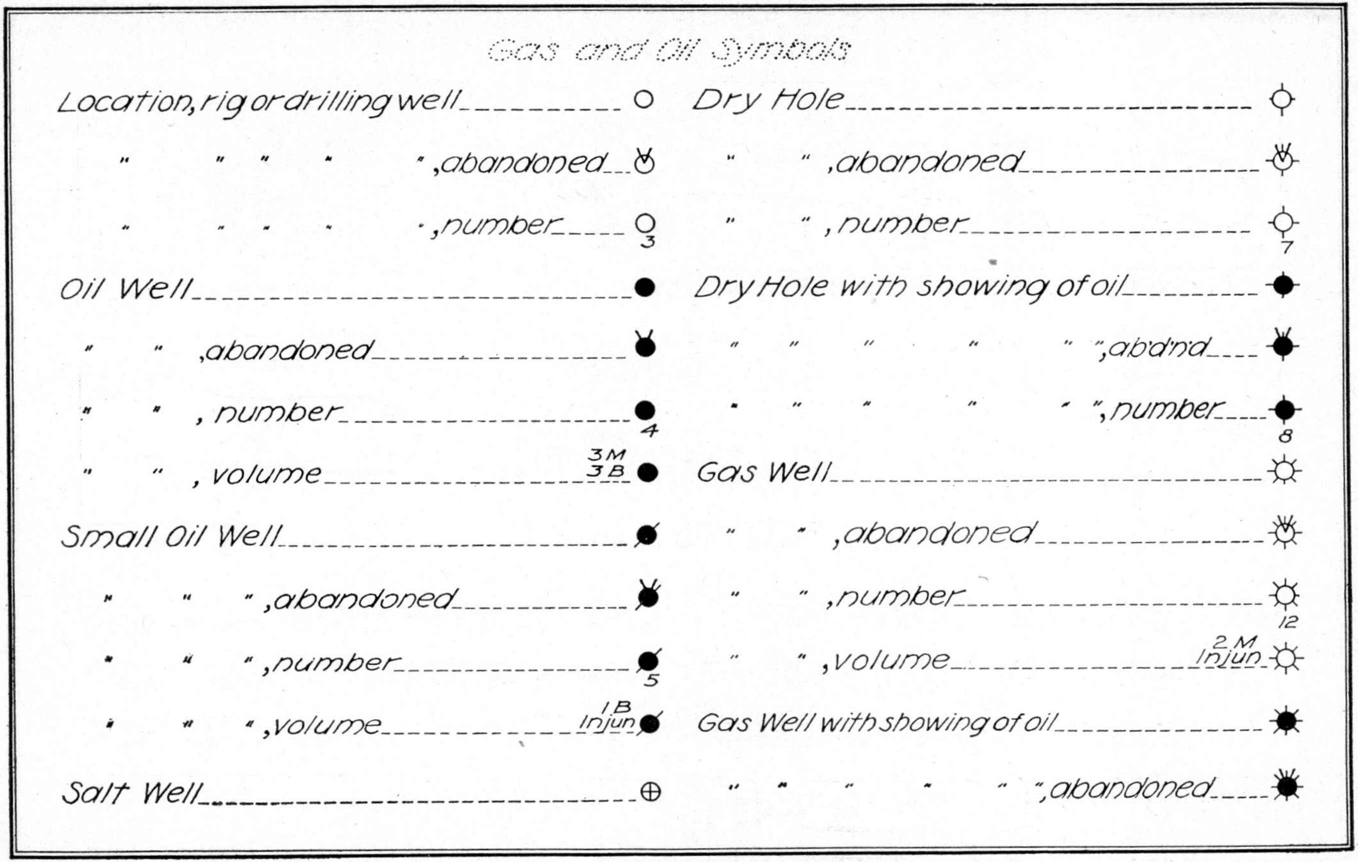

Gas and Oil Symbols

Location, rig or drilling well ⭕	Dry Hole ⌀
" " " " , abandoned ⍉	" " , abandoned ⍉
" " " " , number ⭕₃	" " , number ⌀₇
Oil Well ●	Dry Hole with showing of oil ⬤
" " , abandoned ⍉	" " " " " , abd'nd ⬤
" " , number ●₄	" " " " " , number ⬤₈
" " , volume ³ᴹ/₃ᵦ ●	Gas Well ☼
Small Oil Well ●	" " , abandoned ☼
" " " , abandoned ●	" " , number ☼₁₂
" " " , number ●₅	" " , volume ²ᴹ/Injun ☼
" " " , volume 1.B Injun ●	Gas Well with showing of oil ✹
Salt Well ⊕	" " " " " , abandoned ✹

PLATE 36.

Conventional Signs for Golf Courses

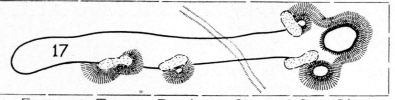

Fairway, Traps, Bunkers, Green, & Open Ditch

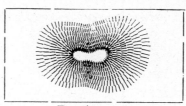

Bunker

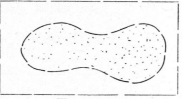

Trap

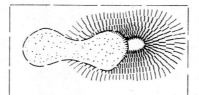

Trap and Bunker

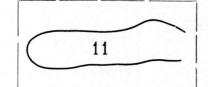

Fairway Outline

Fairway Center Line	Drain Tile	Surface Inlets on D.T.	Outlet on Drain Tile
Sewer Tile	Manholes on S.T.	Sewer Tile Headwall	Irrigation Lines
Hose Outlets	Main Line Valves	Drains	Undergr'd Sprinklers
Drinking Fountain	Fire Hydrants	Drinking Water Line	Tee

PLATE 37

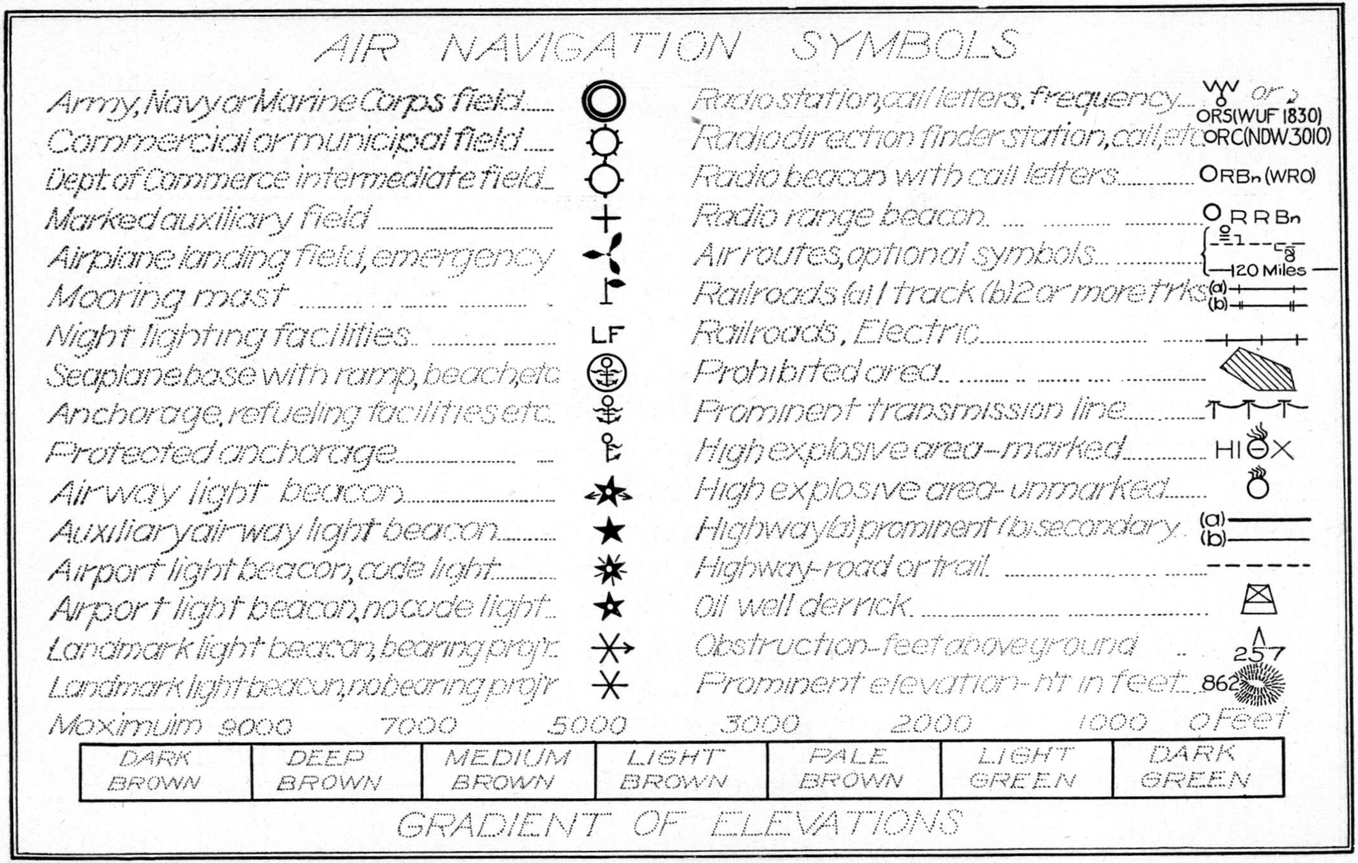

AIR NAVIGATION SYMBOLS

Army, Navy or Marine Corps field	⊚	Radio station, call letters, frequency	⋎⋎ or ⋏ ORS(WUF 1830)
Commercial or municipal field	⌀	Radio direction finder station, call, etc.	ORC(NDW 3010)
Dept. of Commerce intermediate field	○	Radio beacon with call letters	○RBn (WRO)
Marked auxiliary field	+	Radio range beacon	○ R R Bn
Airplane landing field, emergency	🌱	Air routes, optional symbols	⊢—120 Miles—⊣
Mooring mast	🌱	Railroads (a) 1 track (b) 2 or more trks	(a) ─┼─ (b) ═╪═
Night lighting facilities	LF	Railroads, Electric	─◆─
Seaplane base with ramp, beach, etc.	⚓	Prohibited area	▨
Anchorage, refueling facilities etc.	⚓	Prominent transmission line	⊤⌒⊤
Protected anchorage	⚓	High explosive area – marked	HI ⊗
Airway light beacon	✹	High explosive area – unmarked	⬤
Auxiliary airway light beacon	★	Highway (a) prominent (b) secondary	(a) ─── (b) ───
Airport light beacon, code light	✹	Highway – road or trail	─ ─ ─
Airport light beacon, no code light	★	Oil well derrick	⬠
Landmark light beacon, bearing proj'r	✱	Obstruction – feet above ground	△ 257
Landmark light beacon, no bearing proj'r	✱	Prominent elevation – h't in feet	862 ☼

Maximum 9000		7000		5000		3000		2000		1000		0 Feet
DARK BROWN		DEEP BROWN		MEDIUM BROWN		LIGHT BROWN		PALE BROWN		LIGHT GREEN		DARK GREEN

GRADIENT OF ELEVATIONS

PLATE 38.—(*By permission of Federal Board of Surveys and Maps, U. S. A.*)

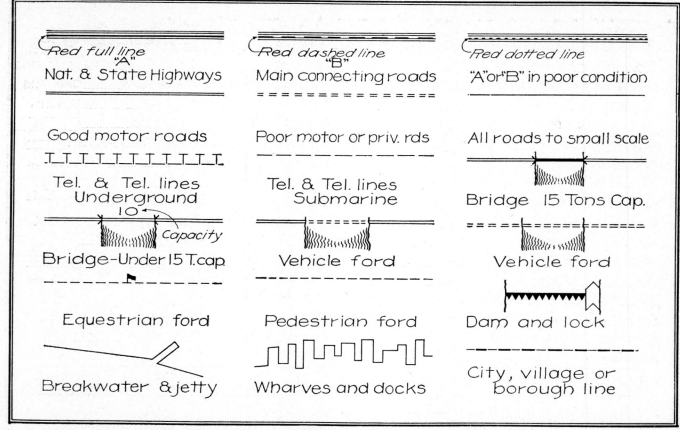

PLATE 39.—(*By permission of Federal Board of Surveys and Maps, U. S. A.*)

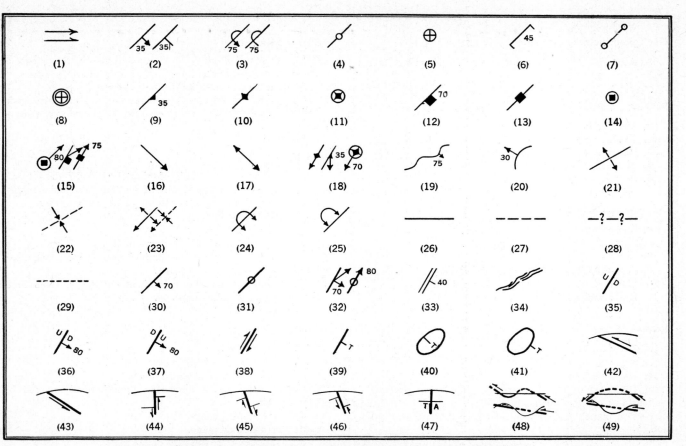

PLATE 40.—Geological-structure symbols. (*By permission of United States Geological Survey.*)

Geologic-structure Symbols.

1. Attitude (double-barbed arrow). Direction (single-barbed arrow).

Bedding.

2. Strike and dip of beds (barbed arrow may be used on the more detailed maps).
3. Strike and dip of overturned beds.
4. Strike of vertical beds.
5. Horizontal beds.

Cleavage and Schistosity.

6. Strike and dip of cleavage of slate.
7. Strike of vertical cleavage of slate.
8. Horizontal cleavage of slate.
9. Strike and dip of schistosity or foliation.
10. Strike of vertical schistosity or foliation.
11. Horizontal schistosity.

Joints.

12. Strike and dip of joint plane.
13. Strike of vertical joint plane.
14. Horizontal joint plane.
15. Direction of linear elements (striations, grooves, or slickensides) on joint planes and degree of pitch of striations on vertical plane. Linear element shown in horizontal projection.

Linear Elements.

16. Direction of pitch of linear parallelism, flow lines, or linear stretching, or alignment of minerals and inclusions.
17. Direction of horizontal linear element.
18. Symbol for linear element may be combined with schistosity symbol as indicated (linear element shown in horizontal projection).

Folds.

19. General strike and dip of minutely folded beds.
20. Direction of pitch of minor folds (anticline and syncline) to determine nature of isoclinal fold at its plunging end.
21. Axis of anticline.
22. Axis of syncline.
23. Pitch of axis of anticline or syncline.
24. Axis of overturned or recumbent anticline, showing direction of inclination of axial plane.
25. Axis of overturned or recumbent syncline, showing direction of inclination of axial plane.

Faults.

26. Known fault.
27. Known fault not accurately located.
28. Hypothetical or doubtful fault.
29. Concealed fault (known or hypothetical) covered by later deposits.
30. Dip of fault plane.
31. Vertical fault plane.

32. Direction of linear elements (striations, grooves, or slickensides, shown by longer arrow) caused by fault movement and degree of pitch of striations on vertical plane.

33. Strike and dip of shear zone.

34. Shear zone (dip of shear plane and direction of striations may be added).

35. U, upthrow, high-angle fault (normal or reverse). D, downthrow, high-angle fault (normal or reverse).

36. Normal fault.

37. Reverse fault.

38. Relative direction of horizontal movement, in shear or tear fault, or flaw.

39. Overthrust low-angle fault. T, overthrust side (opposed to overridden side).

40. Klippe or outlier remnant of low-angle fault plate. T, overthrust side (opposed to overridden side).

41. Window, fenster, or hole in overthrust plate. T, overthrust side (opposed to overridden side).

For Use on Sections

42. Overthrust. Low-angle fault; arrow indicates inferred direction of movement of active block.

43. Underthrust. Low-angle fault; arrow indicates inferred direction of movement of active block.

44. Vertical. High-angle fault (arrow shows relative direction of movement).

45. Normal fault. High-angle fault (arrow shows relative direction of movement).

46. Reverse fault. High-angle fault (arrow shows relative direction of movement).

47. Horizontal movement in shear or tear fault. A, relative movement away from observer; T, toward observer (symbol may be combined with vertical displacement to show diagonal movement).

48. Klippe.

49. Window or fenster.

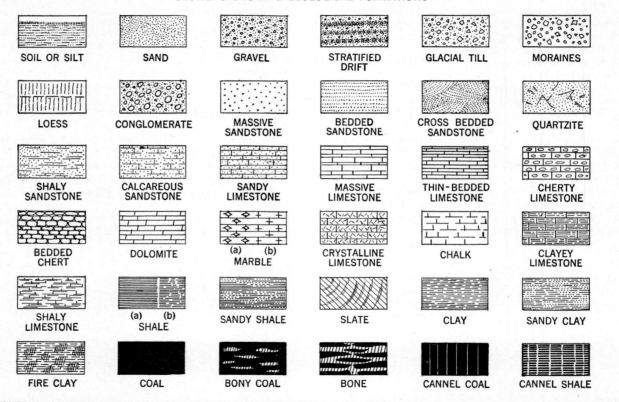

PLATE 41.—(By permission of United States Geological Survey and the Stanford University Press.)

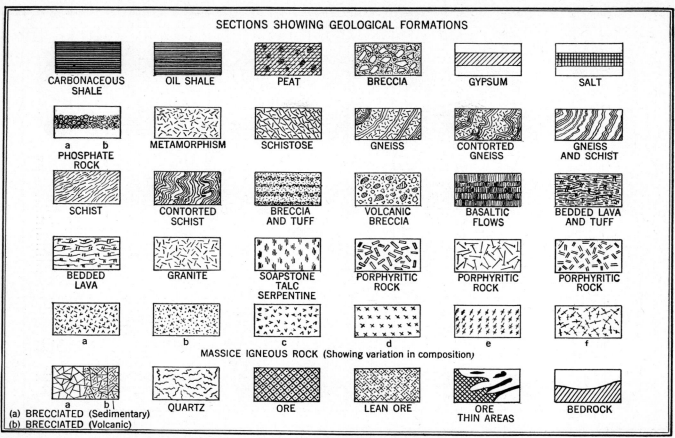

PLATE 42.—(*By permission of United States Geological Survey and the Stanford University Press.*)

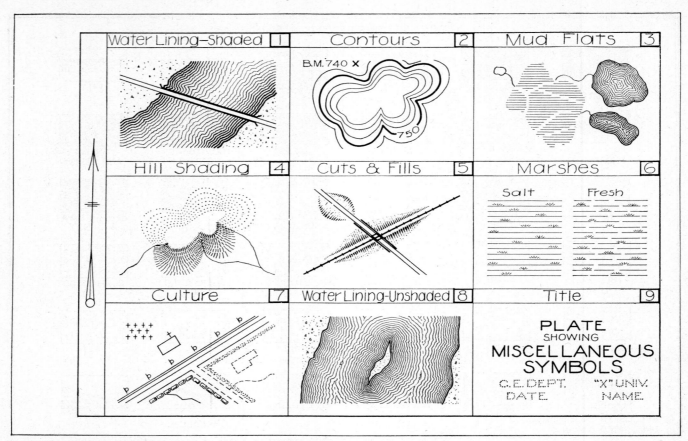

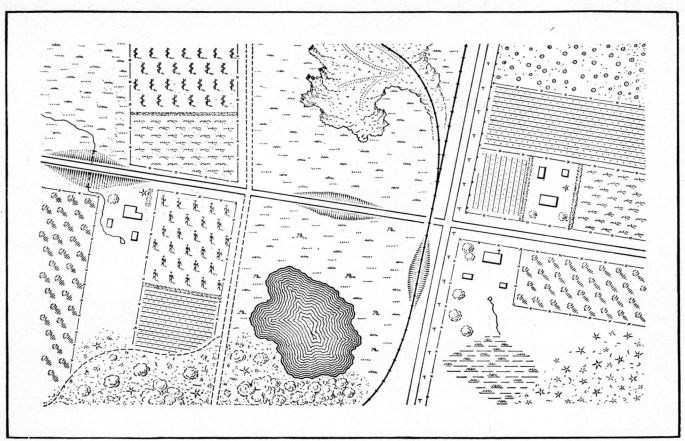

PLATE 44.—Pen and ink topography.

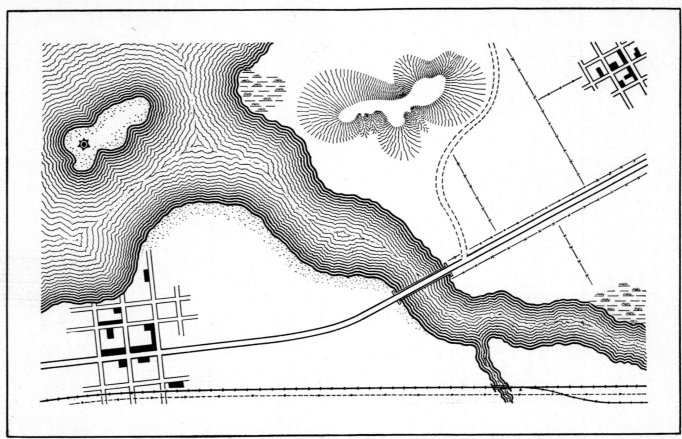

PLATE 45.—Suggested exercise in map toning.

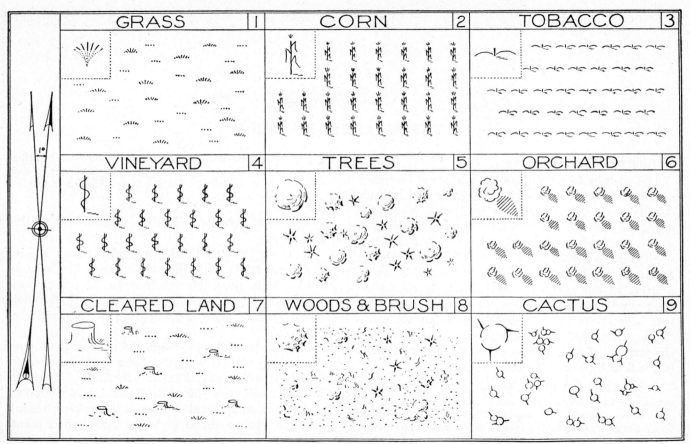

PLATE 46.

CHAPTER III

TOPOGRAPHIC MAPPING

Plotting of Points.

The position of a point on the surface of the earth is said to be known when the longitude and latitude of the point have been determined. When the position of an initial point is either known or assumed, a meridian is drawn through this point and all other points on the survey plotted by their known relation to the initial point. Because of the variation in position of the magnetic meridian it is considered good practice to refer all directions to the true meridian. Directions read from this meridian are called "true azimuths." In topographic survey work, the position of an unknown point may be determined in the field by one of the following ways:

1. By azimuth and distance from a known point, the most common example being transit and stadia surveys.
2. By two angles or directions read from the ends of a known line. Example: Location by intersection with the plane table.
3. By three directions read from the unknown point to three known points already plotted. Example: Two angles read from the boat to three known points on shore as in location of soundings by sextant.

The point thus located in the field may be plotted as follows:

1. Draw through the known point a line making the same angle with the meridian as the azimuth read. Scale along this line from the known towards the unknown point, a distance equal to the field measurement. The result is the plotted position of the point.
2. Draw meridians through the ends of the known line. Lay off angles with these meridians corresponding to the azimuths read in the field. The resulting intersection is the plotted position of the point.
3. From an assumed position of the unknown point, plot the relative directions to the three known points

on tracing cloth. Shift the position of the tracing cloth until the three radiating lines pass through the plotted position of the three known points. The point is now in its true position and can now be plotted by picking through the tracing cloth with a needle or pin.

4. By calculating from the observed field data the coordinates of the unknown point with reference to the assumed coordinate axes through the initial or known point and plotting the point from these axes.

The first and second methods are known as plotting by polar coordinates. The third is the graphic solution of the three-point problem, and the fourth is plotting by rectangular coordinates.

Plotting Traverses.

A traverse is a series of connected lines whose lengths and directions are known. A closed traverse is one enclosing a definite area and having a common point for its beginning and end. An open traverse is one which does not close on the point of beginning. Example: The center line survey for a highway, railroad, etc.

All topographic surveys should have a skeleton or network of traverses to serve as a horizontal control. The stations or instrument points on such traverses should be plotted with accuracy and care, since the correctness of all points plotted from one station depend upon the accurate plotting of the station itself. The following methods of plotting traverses will be discussed in detail.

Plotting by Polar Coordinates.—Figure 19 shows an open traverse plotted by the use of the ordinary 12- or 15-in. paper protractor and engineer's scale. To prepare the protractor for use:

1. Trim the protractor to the neat circle line of graduation in degrees and fractions thereof.

2. Mark graduations from 0 to 360 deg. at intervals of 10 deg.

3. Mount the scale on the circle with its scaling edge along a line connecting the center of the circle with the zero graduation, the zero of the scale being at the exact center of the protractor. The scale may be glued directly to the protractor or connected by a hinged portion of adhesive, as shown in Fig. 19.

4. Cut out a space of $1\frac{1}{2}$ to 2 in. wide along scaling edge of ruler to allow plotting

To plot the line AB, assume the position of A, and draw a meridian through this initial point. When the ruler or zero graduation points north, 180 deg. would read on the south end of the meridian. The

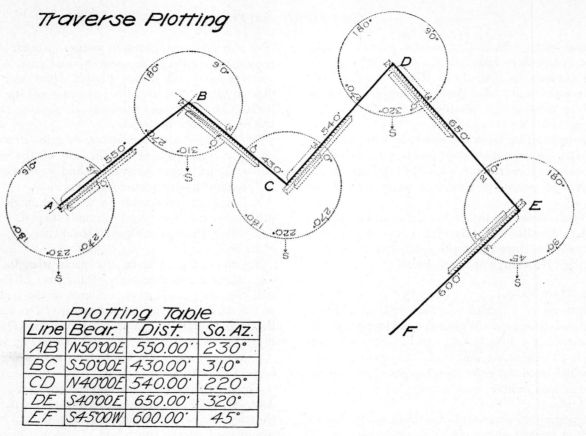

Fig. 19.

protractor circle now occupies the same relative position with regard to a meridian through A as the circle of a transit occupying A in the field survey, when oriented to read south azimuths. To plot line AB turn the protractor until the south azimuth of the line, *i.e.*, 230, is on the south end of the meridian. The ruler now points N. 50 deg.-00 E, and the distance of 550 ft. is scaled on the ruler to locate B. A meridian is now drawn through B and the point occupied by protractor and scale to locate C. Thus, each station becomes a new initial point, until the complete traverse is plotted. The disadvantage of this method lies in the fact that angles cannot be plotted as accurately on paper as they can be read in the field. Plotting errors are therefore cumulative and affect all points subsequently plotted.

To Plot a Traverse by Chords.—Before a traverse can be plotted by chords, a table similar to the one shown in Fig. 20 should be constructed as follows:

1. Columns 1, 2, 3 are taken from the observed field data.
2. The value of the half-angle must be computed by figuring the deflection angle between adjacent bearings and dividing by two.
3. Look up the natural sines of the angles.
4. A radius of 10 units of any scale may be used.
5. The chord lengths are equal to two times the sine of the half-angle, times ten.

The position of the line AB (Fig. 20) having been determined or assumed, the deflection angle at B is plotted as follows:

1. Prolong AB beyond B for 10 units of scale.
2. With B as a center and 10 units of scale as a radius, draw arc xy.
3. With x as a center and the chord length 4.32 units of scale as a radius, draw an arc intersecting xy at y.
4. Connect By for direction of course BC.
5. Locate C by scaling 300 ft. towards y.

Angles at C, D, and E are similarly plotted by using chord lengths of 5.18, 6.84, and 10.00, taken from the table in Fig. 20. The precision attained by this method depends upon the draftsman's skill in setting and using the compasses for describing the above arcs. For arcs as large as 10 in., beam compasses should be used, because they can be more nearly set to the scaled distance and are not likely to spring or slip when being used.

Plotting by Tangents.—The plotting table used in Fig. 21 corresponds very closely to the one used in the chord method, except in the last three columns. The natural tangent of the deflection angle should be used for angles up to 45 deg. For angles greater than 45 deg., use the cotangent. The tangent

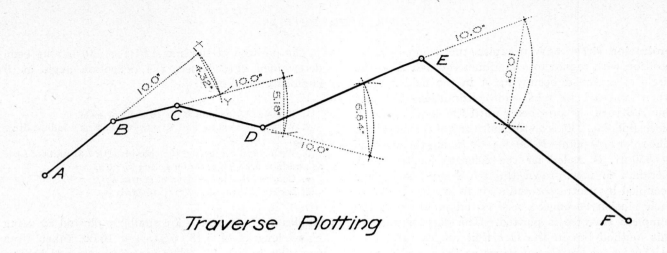

Traverse Plotting

Plotting Table

Line	Bear.	Dist.	Point	½ Angle	Sin.	Rad.	Chord
AB	N30°00'E	400.00	A	0°			
BC	N55°00'E	300.00	B	12°30'	.2164	10"	4.32"
CD	N85°00'E	395.00	C	15°00'	.2588	10"	5.18"
DE	N45°00'E	780.00	D	20°00'	.3420	10"	6.84"
EF	S75°00'E	1200.00	E	30°00'	.5000	10"	10.00"

Fig. 20.

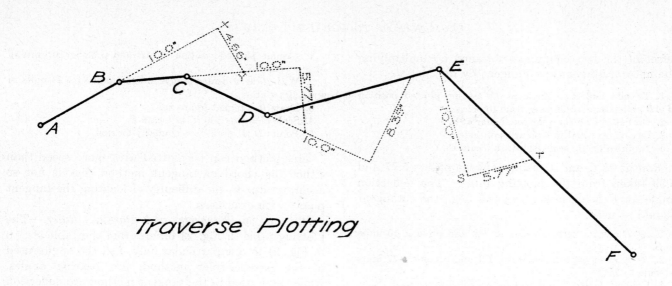

Fig. 21.

Traverse Plotting

Plotting Table

Line	Bear.	Dist.	Point	Angle	Tang.	Cot.	Offset
AB	N30°00'E	400.00'	A	0°			
BC	N55°00'E	300.00'	B	25°R	.4663		4.66
CD	N85°00'E	395.00'	C	30°R	.5773		5.77
DE	N45°00'E	780.00'	D	40°L	.8391		8.39
EF	S75°00'E	1200.00'	E	60°R		.5773	5.77

offset of the last column is obtained by multiplying the natural tangent or cotangent by ten.

1. To plot deflection angle at B, assume plotted position of AB and extend 10 units of scale to point x.
2. On line ABx erect a perpendicular at x.
3. Lay off xy equal to 4.66 units of scale.
4. In direction By scale 300 ft. to locate C.

Repeat at C and D using tangent offsets 5.77 and 8.39 taken from the plotting table. The deflection angle at E being larger than 45 deg., the cotangent should be used.

1. At E erect a perpendicular to DE and prolong 10 units to S.
2. At S erect perpendicular to TS and prolong 5.77 units of scale to T.
3. Connect E and T and prolong scaled distance to F.

Theoretically, the tangent and chord methods should be equivalent in accuracy. Most draftsmen, however, prefer the tangent method, as it is easier to scale distances than describe arcs, if equal degrees of precision are to be maintained.

Plotting by Sines.—Figure 22 represents the same traverse used in the tangent and chord methods, plotted by sines. The calculating of the plotting table is very similar to the chord method and need not be repeated.

1. Assume plotted position of AB and prolong 10 units of scale to point T.
2. With T as a center and a radius equal to 4.23 units of scale, draw arc xy.
3. Draw Bs tangent to arc xy.
4. Scale along Bs 300 ft. to locate C.
5. Points D, E, and F are similarly located.

This method can be plotted with more speed than either the chord or tangent method, but is not so accurate, due to the difficulty of locating the tangent points with exactness.

Plotting by Tangents and Parallel Ruler.—The plotting table in Fig. 23 differs from the table shown in Fig. 21 in one particular only, *i.e.*, the angles used in the parallel-ruler method, are bearing angles, while those used in the tangent method are deflection angles. Calculation of offsets are the same in both methods.

Plot traverse $ABCDEF$ as follows:

1. Select point o near the center of the map.
2. Draw coordinates NS and WE through o as an origin.
3. Enclose the above coordinates in a square having 20 units of scale to a side and the point o for its center.
4. The square is now divided into four quadrants whose sides are 10 units of scale.
5. Using o as an origin, all lines on the survey may be plotted in direction, as shown in Fig. 23.

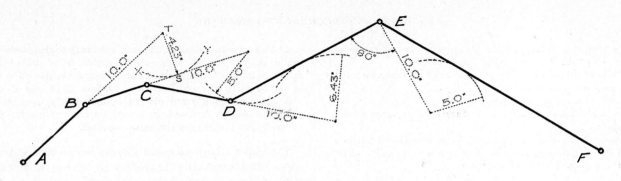

Traverse Plotting

Plotting Table

Line	Bear.	Dist.	Point	Angle	Sin.	Cos.	Radius
AB	N30°00'E	400.00	A	0°			
BC	N55°00'E	300.00	B	25°R	.4226		4.23
CD	N85°00'E	395.00	C	30°R	.5000		5.00
DE	N45°00'E	780.00	D	40°L	.6428		6.43
EF	S75°00'E	1200.00	E	60°R		.5000	5.00

Fig. 22.

6. From this diagram, plot the traverse by assuming a location for AB parallel to its plotted direction.

7. Transfer direction BC by parallel ruler or triangles until its new position passes through B.

8. Scale distance BC to locate C.

9. All other lines on the traverse are similarly transferred and scaled.

10. As a check, the direction AF when transferred parallel to itself should pass through both plotted points A and F. The distance AF should also scale correctly.

Plotting by Coordinates.—The plotting table shown in Fig. 24 is prepared by working out the latitudes and the departures for the entire traverse. The last two columns in the table are obtained by a summation of all latitudes and departures from the initial point A to the point in question. For example, the point F is 1359.79 north and 2299.11 east of A. To plot the traverse, assume the location of initial point A.

1. Take the point A as the origin of coordinates and draw a north-and-south, and east-and-west line through this point. These lines are the zero coordinates.

2. Each point is now plotted by rectangular coordinates referred to the origin A.

3. The sheet is now marked off in squares of convenient size and the various lines marked at the margin of the sheet by their direction and distance from the origin A.

4. To plot a point as at C it is not necessary to scale from the origin A. It can be conveniently scaled from the nearest intersection of coordinate lines at x, *i.e.*, scale 640.08 − 500 or 140.08 north of x and 1022.74 − 1,000 or 22.74 east of x.

5. Since all points are plotted independently, and scaled from the nearest intersection, errors are not cumulative, but are well distributed over the entire traverse.

Plotting by Latitudes and Departures.—Closed traverses, and especially traverses involving a calculation of area, are very conveniently plotted by the above method. The plotting table consists of the figured latitudes and departures for the traverse, as shown in Fig. 25. To plot the survey, proceed as follows:

1. Assume plotted position of A and draw a meridian through this point.

2. All latitudes will be parallel to this meridian and all departures perpendicular to it.

3. Care must be taken to preserve the above parallelism and perpendicularity or the accuracy of the plotting is impaired.

4. To plot point B, it will be observed from the table that the direction of AB is northeast, its northing being 400 ft. and its easting 200 ft. Scale north from A along the meridian 400 ft. to x. Erect perpendicular at x and scale 200 ft. east to B.

5. Connect A and B by line AB and check by scaling AB equal to the tabular value given as 447.24 ft.

6. To plot point C, scale east from B 400 ft. to y. Draw a line through y parallel to meridian through A and scale 150

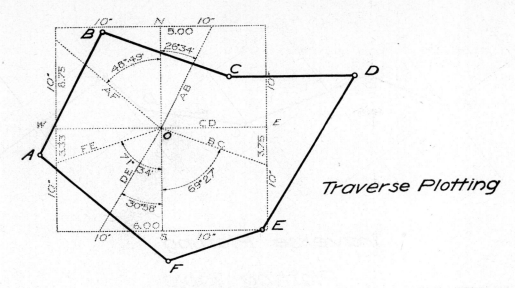

Traverse Plotting

Plotting Table

Line	Bear.	Dist.	Angle	Tan.	Cot.	Offset
AB	N26°34'E	447.24	26°34'	.5000		5.00
BC	S69°27'E	427.17	69°27'		.3749	3.75
CD	EAST	400.00	90°00'			0.00
DE	S30°58'W	583.09	30°58'	.6001		6.00
EF	S71°34'W	316.22	71°34'		.3333	3.33
FA	N48°49'W	531.49	48°49'		.8749	8.75

FIG. 23.

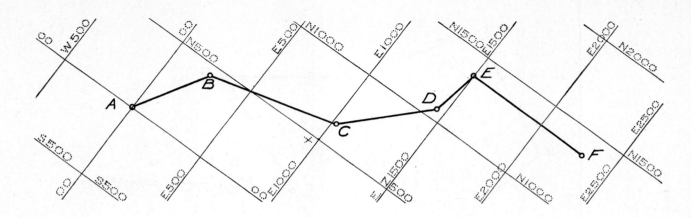

Traverse Plotting

Plotting Table

Line	Bear.	Dist.	Sin.	Cos.	N	S	E	W	SumLat.	SumDept.
AB	N30°00'E	500.00	.5000	.8660	433.02		250.00		N433.02	E250.00
BC	N75°00'E	800.00	.9659	.2588	207.06		772.74		N640.08	E1022.74
CD	N45°00'E	600.00	.7071	.7071	424.27		424.27		N1064.35	E1447.01
DE	N10°00'E	300.00	.1736	.9848	295.44		52.10		N1359.79	E1499.11
EF	EAST	800.00	1.0000	.0000	0.0		800.00		N1359.79	E2299.11

Fig. 24.

ft. south from y to C. Scale BC and compare with tabular value.

7. Proceed around the survey in this manner until point F is located.

8. If plotting has been accurate, AF should scale very close to its tabular value of 531.49 ft. Fz should scale 400 ft. and Az 350 ft.

Hints and Precautions.—When the deflection angle between adjacent courses is small, sizeable errors may be introduced into the plotting by the method of sines, tangents, or chords, since the scaled distance is small and the possible error of scaling remains the same.

When the angle between any course and the reference meridian is small, considerable error may be made in plotting the total departure of either end of the line without materially changing its scaled length. Such courses should have the additional check of scaling the respective latitude and departure, and comparing same to calculated results.

A good check, when plotting by balanced latitudes and departures, is to lay off a rectangle equal in dimensions to the total latitude and departure. The sides of the above rectangle being north-and-south, east-and-west lines, each side should have at least one plotted point fall directly on the line. Start the plotting by placing the westernmost point on the survey in its correct position as to latitude on the west side of the rectangle. Proceed to plot the traverse by latitude and departures until the northernmost point of the survey is plotted. This point should fall on the north boundary of the rectangle. Failure to do so shows the accumulated error in latitudes and the drawing should be rescaled and corrected. Similarly, a check on departures is secured when the easternmost point on the traverse is plotted. Thus, each time a plotted point falls on the perimeter of the plotted rectangle, a check on either latitudes or departures is secured.

The coordinate method is recognized as the most reliable and foolproof way of plotting traverses. Its one great disadvantage is the large amount of calculation preliminary to plotting. In case area calculations are required, the preliminary labor of figuring total coordinates is little more than that required by tangents or chords.

The coordinate method has the following advantages:

1. For closed traverses, the field work is checked and survey balanced before plotting is begun.

2. Method of checking by scaled length of course is simple and does not involve a repetition of scaling plotted latitudes and departures.

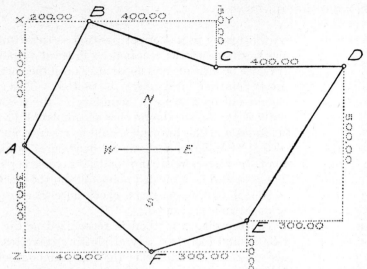

Traverse Plotting

Plotting Table

Line	Bear.	Dist.	Sin.	Cos.	N	S	E	W
AB	N26°34'E	447.24	.4472	.8944	400.00		200.00	
BC	S69°27'E	427.17	.9364	.3510		150.00	400.00	
CD	EAST	400.00	1.0000	.0000			400.00	
DE	S30°58'W	583.09	.5145	.8575		500.00		300.00
EF	S71°34'W	316.22	.9487	.3162		100.00		300.00
FA	N48°49'W	531.49	.7526	.6585	350.00			400.00

Fig. 25.

3. The accuracy of the plotted position of a point does not depend upon the accuracy of points previously plotted.

4. The effect of accumulated plotting errors is eliminated.

5. The size and shape of the proposed drawing can be determined before plotting is begun.

Blocking Out of Map.

Before any of the detail work is done the map should be roughly outlined in pencil:

1. Decide on scale to be used.

2. From the field notes and traverse calculations determine size and shape of area to be mapped.

3. Lay out borders and allot space for title, north point, legend, and explanatory notes.

4. Arrange above outline symmetrical to border line.

5. Arrange space for important lettering as otherwise it may be obscured by detail.

6. Follow some general order of plotting, *viz.*, buildings, communications, streams, enclosures, marshes, woods, and remaining features.

7. Contours should be plotted last.

8. Use a sharp 4H or 6H pencil and draw **very** light lines.

9. Keep paper clean and avoid erasures.

Plotting of Details.

The methods used in plotting details on the map **correspond** very closely with the methods used in the field for locating the above details. This is entirely reasonable, since the draftsman is representing to a reduced scale the actual horizontal distances and angles read in the field. Since the field work in locating details is less refined than the traverse or skeleton survey, the details are not plotted with the refinement and care used for the traverse or horizontal control. In general, the aim is to plot objects of definite size and shape so that their plotted positions will scale correctly within the allowable error of the map.

Plotting Stadia Notes.

If a considerable number of points are to be located from an instrument station, some form of protractor adapted to laying off both angle and distance should be used. Paper protractors graduated from 0 to 360 deg., and fitted with a distance scale, are in general use. The protractors shown in Figs. 26, 27, and 28, can be purchased at any engineering-supply store and fitted to the scale of the map being made.

Figure 26 shows a protractor to be fitted with a scale and numbered to plot south azimuths. The scale has its zero point at the center of the graduated circle, while its scaling edge connects the center of the circle with the 0-deg. graduation. An annular space

104   ELEMENTS OF TOPOGRAPHIC DRAWING

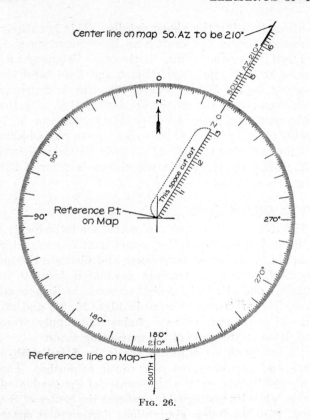

Fig. 26.

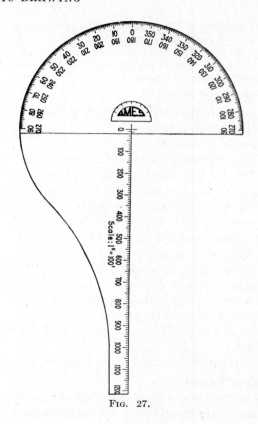

Fig. 27.

about 2 in. wide is cut out of the protractor circle along the scaling edge of the rule to allow plotting space. The rule is either glued to the protractor circle or connected by a strip of adhesive tape, to avoid covering up the circle graduations. To plot an azimuth of 217 deg. 15 min. and a distance of 300 ft., proceed as follows:

1. Draw a meridian through the instrument point to be occupied by the protractor, and extend north and south, a distance of 1 in. beyond the graduated circle.
2. Fasten the center of the circle to instrument point on map with a needle or pin about which it can rotate.
3. When the ruler points north along the meridian, the 180 deg. is at the south point and the 0-deg. graduation at the north point.
4. To plot the given point, revolve the ruler clockwise until 217 deg. 15 min. is opposite the south point, scale 300 ft. along ruler and make point.

It will be noted that where the protractor and scale move as a unit, the circle should be graduated counter-clockwise to orient the ruler properly. Figures 26 and 27 show protractors of this type. In Fig. 28 the circle is graduated clockwise because the protractor remains stationary with the 180-deg. graduation on the north meridian. The ruler is then revolved to the required azimuth, since the protractor is always oriented.

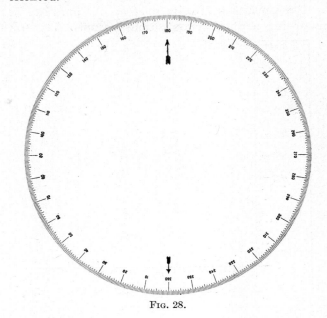

Fig. 28.

Use of Two Protractors.

When points are located in the field by two azimuths from opposite ends of a known base, the plotted position lies at the intersection of the two azimuth lines.

If the points are few in number, the lines can be numbered and the two stations occupied in turn.

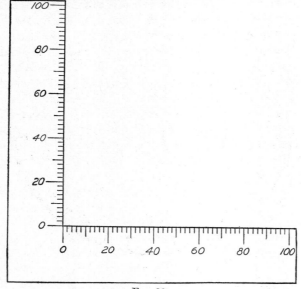

Fig. 29.

When a large number of points are located from one base, as soundings in coast-line work, it will speed up the work of plotting to use a protractor on each end of the plotted position of the base line. These protractors should be smaller than the ordinary paper protractor and graduated to suit the field observations.

Rectangular Protractor.

When details are located by rectangular coordinates, as in the so-called "checkerboard surveys," much labor can be saved and many possible errors eliminated by constructing a rectangular protractor from heavy pasteboard. Cut out a piece of pasteboard in the shape of a carpenter's square, as shown in Fig. 29. The width of the legs should be about 1 in., and the length equal to the distance between coordinates. From the inside angle of the square, graduate both ways to the scale of the map. To plot a point referred to any intersection of coordinates, place one leg of the protractor along the east-and-west coordinate, with the scale reading the east or west measurement at the intersection. Follow along the vertical leg to the scale reading of the north or south measurement to the plotted position of the point.

Checking.

The location of all important details should be checked by actual map measurements before being inked. Less important details are checked by observing their relation to known points on the map. Mistakes in field work often give two locations for the same point. If this discrepancy cannot be removed by the above inspection, the points in question should be relocated in the field.

Penciling Details.

Symbols covering areas such as orchards, woods, cultivated land, crops, swamps, sand bars, mud flats, ponds and lakes, should not be penciled in detail. The boundaries of such areas located in the field should be carefully plotted and shown by light pencil lines, together with the name of the symbol covering the area written lightly within the area. This procedure not only saves much valuable time but prevents mistakes in shading and execution resulting from the inker trying to retrace the penciled symbols. If the orchard to be represented has been actually located as to direction of rows and spacing of trees in the above rows, the same should be shown on the map. Locate the plotted position of the grid system, and make a distinct dot for the location of each tree. In inking, build the tree around the dot as a center.

Trees.—In large-scale and landscape work, the position, kind, and size of each tree is recorded in the field notes and should be plotted in pencil by making a dot for the plotted position and indicating species and size by abbreviations. This enables the inker to use the proper symbol for each tree and maintain a scale of relative sizes.

Fences.—The plotted position of fence lines should be penciled as a light, full line. The kind of fence as barbed-wire, smooth-wire, hedge, stone, picket, board, or worm should be written along the line so that the proper symbol may be used in inking.

Roads and Trails.—All roads, trails, boulevards, streets, bridges, and tunnels should be penciled as final copy, for inking according to the distinction required.

Houses.—In large-scale topography, houses are penciled in full outline of their ground projection. If a distinctive symbol is to be used for frame, stone, or brick, the word should be abbreviated and written near the building as a guide for inking. The conventional black square is used ordinarily to represent buildings in small-scale topography. Whenever the dimensions of the building, when plotted to scale, exceed the size of the symbol, the actual plan dimensions of the building should be shown. Churches are to be distinguished by a cross, and schoolhouses by a pennant, so attached to the house symbol as to point at right angles to the roadway.

Railroads.—All railroads, whether operated by steam, electricity, gasoline, or other motive power, should be penciled by a single full line showing its center-line location. The name of the railway, together with its classification, should be written along the center line in pencil.

Grade crossings should be penciled showing both railroad and highway symbols continuous; a railroad crossing over a highway by breaking the road symbol; and a highway passing over a railroad by breaking the railroad symbol.

The symbol showing the classification of both railroads and highways should be omitted in penciling and applied when inked.

Relief Sketching.

Cuts, fills, levees, and eroded banks intended to be shown by hachures or sketched in oblique should be lightly outlined in pencil as a guide to inking. The above should never be penciled in full, as they are difficult to trace in ink without giving a stiff and cramped appearance.

Drainage.

Care should be taken to show small and intermittent streams, ridge lines, summits, and depressions, with plainly penciled lines in their true plotted positions, but no attempts should be made to water line in pencil, as this is a detriment rather than an aid in inking.

Penciling Contours.

Contour lines should be drawn with a relatively hard, sharp pencil. The pencil lines should be fine and of even strength, but under no circumstance should the pencil be allowed to dent the paper or form grooves in its surface.

Drainage Lines.—All stream and ridge lines should be plainly shown, together with the field elevations which determine them. The above lines serve as a natural skeleton for the proper construction of the contours, and are invaluable to the topographer in making clear the real nature of irregular eroded slopes and surfaces.

Emphasized Contours.—Every fourth or fifth contour, depending upon the contour interval, is emphasized by inking it in a heavy line in order to more forcefully bring out the relief. Such contours should be marked by light penciled crosses, and given their respective numbers, as a guide in inking. Where the slope is both steep and uniform, only the accentuated contours should be penciled, for the reason that the interpolated contours can be drawn just as easily when inking, and the penciled sheet is much clearer. Where the slope is steep but not uniform, intermediate contours should be penciled to define clearly the positions of the changes in slope.

Stream Crossings.—Much of the beauty and effectiveness of a contour map may be destroyed by the manner in which contours are drawn across streams or gullies. In rugged country where the stream banks are steep and the fall of the stream is great, contours should be close together near the stream, travel sharply up stream, and cross the watercourse at right angles. Contours crossing a stream having a wide valley and little fall per mile would approach the stream at a greater angle, bend slightly upstream and cross at right angles.

Depressions and Sinks.—Depressions occurring in the limestone regions, artificial enclosed embankments, and other depressions should be marked by hachures on the downhill side of the contours enclosing them. Large depressions shown by contours should be drawn without hachures, provided there is ample room to number the contours without crowding.

Cuts and Fills.—Slopes trimmed to grade, when designated by contours, should be drawn by straight penciled lines rather than drafted freehand. If contours are crowded too closely, resort to hachures or oblique sketching to represent the feature.

Cliffs.—In representing cliffs and outcropping rock formation, a number of contours may be merged to form a wide line or band, and the individual contour would then lose its identity. In such cases, number enough contours in the vicinity of the cliff to ensure correct reading, and resort to hachures or rock sketching to represent the cliff, as shown in Plate 10.

Hachures.

The use of hachures should be confined to features which cannot be shown by contours, yet are of such a degree of importance as to prevent their omission. Banks or levees occurring between contours, and of less height than the contour interval, should be shown by hachures. Small depressions and mine dumps having steep sloping sides are very difficult to contour and in general are shown either by hachures or oblique sketching.

Inking a Topographic Map.

A topographic map should be so made that its features appear in the order of their importance when viewed at the ordinary map-reading distance, *i.e.*, 14 to 20 in. from the eye. The most important features, as buildings, roads, streams, and accented contours appear at first glance, thus giving the observer a mental picture of the general culture and relief.

Features of secondary importance, such as intermediate contours, fences, shrubbery, grass, etc., appear as added detail to the first mental picture. The details for inking the various symbols, hill shading, water lining, contouring, etc., are given in the chapter on Conventional Signs, and need not be repeated here.

In inking a topographic map, the draftsman may have at most but few and minor erasures, and should so plan and execute his work that erasures are unnecessary. This can be accomplished only by having some definite order of precedence in inking the various classes of information. All draftsmen do not follow the same order of precedence, yet there are certain fundamental principles which must be used as a guide, to avoid confusion.

1. Certain classes of symbols, such as towns, houses, railroads, roads, fences, etc., have no option as to location.

2. To complete a map, one class of information must be superimposed upon another without unnecessary interference in showing a clear record.

3. The draftsman must keep clearly in mind the degree of importance it is desirable to give each topographic feature on the finished map. This naturally suggests the inking of symbols in the order of their importance.

4. The element of speed demands some fixed **routine** or order of work.

Hints and Precautions on Order of Inking.

1. Towns, villages, and houses should be inked first; primarily because of the importance of this class of information and because they have no option in location. Since they constitute the heaviest and blackest symbol on the map, their number and disposition largely fix the degree of blackness or scale of shading on the finished map.

2. Roads, railroads, and lines of communication should be inked second. This class of information also has no option as to location. Care must be taken as to order of precedence where lines of communication cross, as explained under details of penciling.

3. All lines of authority or ownership, together with fences and lines showing rights of way, are shown next because the symbol is continuous and has no option of location.

4. All names should be lettered in ink at this time, as they have precedence over all information following. Use care to see that the lettering does not obscure other information.

5. The numbering of contours having been studied and penciled, the contours may now be inked. Ink all accentuated contours before inking intermediate ones. Do not draw contours through houses or across roads, as no additional information is thus given, and the road symbol, especially in rugged country, is obscured.

6. The inking of the contour numbers is done after the inking of the contours themselves, in order to ensure the figures being on the exact axis of the contour.

7. Ink all water features next, following the rules for water lining given under Conventional Signs. Use care in drawing the small single-line streams, to see that they cross the contours accurately at right angles at points representing the lowest part of the valley. This is exactly the reverse order to penciling contours and streams. In penciling, the stream is drawn first, and the contours made to fit the stream.

8. Since the vegetation symbols are subordinated to all other classes of information, they should be drawn last. The vegetation symbols should have an order of importance among themselves. Trees, especially in landscape work, should have a relative scale of sizes for inking. Shrubbery should be inked lighter and finer textured than trees. Hedges should be represented by light, irregular lines, and should not resemble a row of individual trees. Crop symbols should never be so heavy as to obscure contours or unbalance the map. The symbols for grass or cleared land should be used sparingly where trees or shrubbery are plentiful, and more generously in the open white spaces on the map. Do not draw this symbol in rows, and avoid tall scratchy lines. Many draftsmen ruin an otherwise good map by careless use of the grass symbol.

9. After the map is otherwise finished, the title is added according to the instructions given under Map Lettering.

Lettering for Topographic Maps.

The type of lettering selected for topographic maps is more or less controlled by convention and is simple and dignified rather than ornamental. Civil divisions and hydrography are lettered in upright modern Roman, and in inclined modern Roman, respectively, the most important features being lettered in capitals and the less important in lower-case letters. Hypsography, or land features, are lettered in upright commercial Gothic, both upper- and lower-case, according to importance. Public works and culture are lettered in inclined commercial Gothic, upper- and lower-case, according to importance.

It is assumed that the student taking up topographic drawing has already had a course in mechanical drawing and lettering, and is therefore familiar with the use and care of drawing instruments and has some knowledge of lettering. He will, however, find it necessary to go into the subject of lettering much more in detail.

Convention demands that certain names shall be recorded in capitals, while others are recorded in lower-case letters, and finally, that different classes of information be recorded in different alphabets. The topographic draftsman must master the above rules, acquire a knowledge of several types of alphabets, and develop the ability to execute them according to a high standard of excellence.

Since in making a topographic drawing a high standard of freehand draftsmanship is desirable, the lettering on these maps should be of a distinctive character and merge harmoniously with the rest of the drawing. The draftsman must develop skill and judgment in the size and distribution of map lettering; therefore, a few simple rules for the beginner may not be amiss:

1. Names should be so placed as to be easily read and show clearly the object designated. They should not obscure topographic symbols.

2. Place the names of isolated objects, where possible, to the right or the left of the object designated and as close to the object as practicable.

3. Names should be parallel to the lower border, the letters and words spaced as in ordinary print.

4. Railroads, roads, and other communications should be lettered parallel and close to their right-of-way boundary. The letters in the words have the ordinary spacing, while the words themselves are separated by a space equal to the longest word in the designation.

5. Letter with the bottom of the letters nearest the communication and select a place where the lettering reads from the bottom of the map, *i.e.*, where it will not be reversed.

6. Streams follow the same rule as for communications, except where the width of the stream is at least twice the height of the letter used; in which case, letter the name on the axis of the stream previous to water lining.

7. In forests, swamps, and bodies of water having elongated outlines, the names should be extended in the direction of the longest dimension, either in straight or slightly curved lines occupying the center of the tract. The letter spacing should not be extended, but the words should be separated until the name extends almost the entire length of the feature.

8. In case the scale of the map is too small to designate the name, as in No. 7, the same should be included in a legend placed at an important place on the map or entirely outside its limits.

9. Make a study of sizes and proportion all lettering to suit the size of the map and the relative importance of the feature it represents.

10. All names lettered in lower-case letters have capital initials.

11. The names of states, counties, townships, etc., should be lettered in the center of the area represented. The lettering should be parallel to the bottom border. If not, letter in slightly curved lines in center of area.

12. Do not attempt to letter without guide lines and be careful to keep tops and bottoms of all letters exactly on the guide lines. Rounded letters such as O, C, S, Q, and G should extend about one-half a hair-line width above and below the guide lines.

13. Omit fillets from Roman letters less than $3/8$ in. high, as they are difficult to draw in proportion to the letter.

14. Keep the apparent black and white spaces in each word well balanced.

15. The tops of capitals E and Z are narrower than their bases. Failure to observe this gives the letter a top-heavy appearance.

16. The top portions of capitals B and S and the figures 3 and 8 are always smaller than the lower half.

Size of Letters.

The height of letters used on topographic drawings should be proportionate to the size of the drawing. In no case should they be made so prominent as to obscure the more important symbols shown.

Having determined the proper height of letters for a given drawing, the following rules for determining the approximate size of capital Roman and Gothic letters are recommended:

a. Make the stems of all letters one-sixth the height in thickness. The tendency is to make them thicker than this.

b. Considering any given height as divided into six equal units, letter widths are as shown in the following table.

Examples of detailed letter construction are shown in Figs. 30 and 32 for three representative capital Roman and capital Gothic letters. Again it should be emphasized that the over-all bottom widths of capital letters B, E, G, K, R, S, X, and Z are made slightly wider than the top widths in order to avoid a top-heavy appearance. This is also true of numerals 2, 3, 5, and 8. Lower-case letters should be made three-fifths the height of their corresponding capitals.

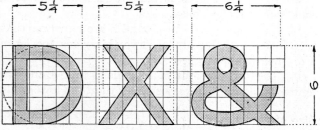

Fig. 30.—Examples of letter construction

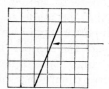

Fig. 31.—Slope of slanting letters.

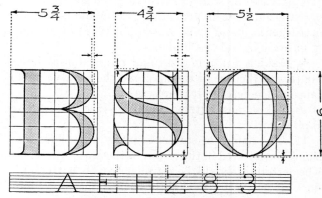

Fig. 32.—Examples of letter construction.

Letter	Units in Width Over-all
I	1
J	3¾
U, N	4¼
F, H, L, P, R	4½
B, E, S	4¾
A, T, V, Y, Z	5
C, D, G, X	5¼
K, O, Q	5½
M	6
W	7½

Inclination of Letters.

The slope of inclined letters should be about 70 deg. to the horizontal or a ratio of two horizontal to five vertical, as shown in Fig. 31. It is best, particularly for beginners, to draw in guide lines at this inclination, before attempting to put in the letters.

Types of Letters Used.

Civil and Political Divisions.

1. The names of states, counties, townships, capitals, and large cities should be in upright modern Roman capitals, the relative size of the letter used depending directly upon the scale of the map and the position in the scale of importance of the particular feature named. The name of the most important political unit within the area mapped is shown in the largest letters of this type, and minor subdivisions scale down from the above size, as shown in Fig. 33.

2. The names of small towns, villages, post offices etc., should be shown in upright, lower-case modern Roman, with the exception that the first letter in each word should be a capital.

OHIO, FRANKLIN CO. PERRY TWP.

FIG. 33.

3. Maps having many towns varying in size and importance call for a gradation in scale in lettering their names, as, for example, in Fig. 34.

4. In no case, however, shall the lettering be prominent enough to obscure the topography or small

Browntown .   Lucasville

FIG. 34.

enough to be illegible. The name should be lettered parallel to the bottom border and may be located anywhere, so long as the name refers unmistakably to the correct place. The positions in order of their preference are, to the right, to the left, centrally below, and centrally above.

Hydrography.

1. Names of oceans, bays, gulfs, sounds, large lakes, and rivers are lettered in inclined modern Roman capitals. The sizes of the letters are in proportion to the scale of the map and the importance of the feature shown as in political divisions (see Fig. 35).

ATLANTIC OCEAN
LAKE ERIE
Fig. 35.

2. The names of small rivers, creeks, lakes, ponds, marshes, brooks, and springs should be shown in inclined modern Roman lower-case or stump letters. The first letter in each word should be capitalized as, for example in Fig. 36.

Spruce Lake, Turkey Run
Fig. 36.

The lettering of the hydrographic features on a map is a severe test of the skill and artistry of the draftsman. The winding courses of streams offer many problems in the spacing and arrangement of lettering. It is well to neglect the lesser sinuosities and sketch curved guide lines parallel to the larger bends of the stream. Normals to the guide lines should be sketched lightly where the letters are to be placed, and the whole name blocked out in pencil. Check the height of each letter, and, if the arrangement is satisfactory, proceed with the inking. If possible, letter the names of lakes entirely within its shore line; if not, letter the name entirely outside the shore line and parallel to the bottom border, follow the same order of preferred position as in the names of towns.

Hypsography.

1. Prominent natural land features, such as mountain ranges, plateaus, canyons, valleys, etc., should

BLUE RIDGE MTS. PIKE'S PEAK
Fig. 37.

be shown in upright commercial Gothic. These names are subordinated to the names of civil divisions shown on the same map, hence are reduced in size, the largest size being smaller than the lower scale in Fig. 33. The lettering of the various hypsographic symbols is graded in size according to the relative importance of the feature represented, as, for example, in Fig. 37.

2. The names of small valleys, peaks, islands, ridges, etc., are lettered in lower-case, upright commercial Gothic, with capital initials. The variation in size of letters is at all times governed by the importance of the feature names, as, for example, in Fig. 38.

Chestnut Knob   Scott's Ridge

Fig. 38.

Public Works.

The names of railroads, highways, tunnels, bridges, ferries, trails, fords, and dams are lettered in inclined Gothic capitals. Since the features named are relatively frequent on the majority of maps made, the

PENNSYLVANIA R.R.   REFUGEE ROAD

Fig. 38a.

lettering is subordinated to other features. Being considerably smaller in size, it is fairly uniform for all names included in this subdivision (see Fig. 38a).

Map Titles.

Topographic maps offer a special study in the subject of titles, and no definite set of rules can be given to fit all cases.

In general, the title is proportioned to the size of the drawing and is placed in the lower right-hand corner sufficiently far from the border to give it a proper margin of white. The wording of the title should first be written out and a study made of the relative importance of each line. The line conveying the most important information should be made the major line of the title and is drawn in the largest letters to emphasize this fact. The remaining lines are lettered smaller in proportion to their importance.

The various lines constituting the complete title should be symmetrical with respect to a center line or axis drawn perpendicular to the bottom border. Count the letter spaces in each line, allowing one space between words and one space for each letter. This will give you the approximate center of each line as a guide in lettering. Block out the lettering in pencil, starting with the letter forming the center of the line lettered on the axis of the title. Work both ways from the center until the complete line is lettered, and check the distances from the axis to the two ends for symmetry.

Use care in the spacing between the various lines of the title to give it a neat, well-balanced appearance. The types of lettering used should correspond to the alphabets used in the body of the drawing, *i.e.*, a topographic map using modern Roman letters for important civil divisions and Roman lower-case for secondary work should have the main lines of the title lettered in upright modern Roman capitals. The scale, date, surveyor's name, and other minor details

MAP OF OHIO
SHOWING
PRINCIPAL STREAMS
AND THEIR
DRAINAGE AREAS
C. E. SHERMAN, C. E.
JULY, 1925

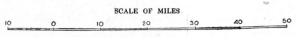

SCALE OF MILES

FIG. 39.

should be in upright lower-case Roman (see Fig. 39). A map showing a proposed building site lettered in single-stroke commercial Gothic and single-stroke lower-case should have a title lettered in the same alphabets (see Fig. 40).

All title lettering on topographic maps should be vertical, for the following reasons:

a. Vertical lettering presents a more dignified appearance.
b. It has a proper relation to the border lines.

Some forms of maps, notably the quadrangle sheets of the U. S. Geological Survey, do not use the symmetrical title because the entire area is occupied

McGUFFEY SCHOOL		
THE BOARD OF EDUCATION COLUMBUS OHIO		
DR. BY. HK	HOWARD DWIGHT SMITH ARCHITECT	DATE 2-1-1928
SCALE 1/30 = 1'-0"	SITE IMPROVEMENT	DRWG No 2-S

FIG. 40.

by the mapped details. Such maps use the marginal or split title, having the information recorded at various places around the border margins. The form of lettering used is the light-faced Gothic shown on Plate 50.

The amount of information contained in the title varies with the draftsman's dislike for particular lettering and his conception of what a title should contain. All maps do not call for the same amount

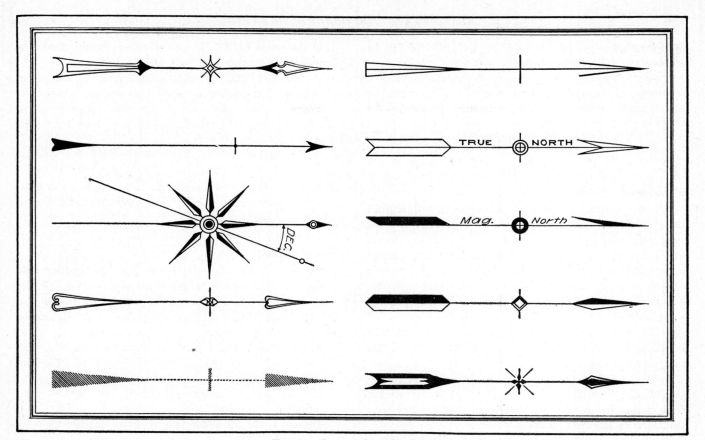

Fig. 41.—Suggested north points.

of information in the title, yet no title should fail to convey the following details:

1. The kind of drawing: topographic, profile, etc.
2. Where the area mapped is located.
3. What the drawing is primarily intended to show, as for example, "Columbus Reservoir Site," etc.
4. In what office the drawing was made.
5. The draftsman's and the checker's names.
6. The scale used.
7. The date on which the drawing was completed.

Meridian or North Point.

All topographic maps, no matter how small, should have a north point showing the direction of true or magnetic north. Magnetic north is shown by a half arrow and true north by a full arrow. Since many draftsmen ruin an otherwise good map by their choice of position and design for a north point, the observance of the following simple rules may be of some help:

1. In blocking out the map, orient it so that north is either at the top or at the right-hand side of the map—preferably the former.
2. The north point should be proportionate to the size of the drawing. For ordinary maps, its length should not be over one-fourth the distance between the borders towards which it points.
3. If the drawing is long compared with its width, and the north point is parallel to its length, the above rule does not hold. The draftsman should choose a length that harmonizes with the drawing.
4. Do not make north points ornamental. The simpler and the more dignified, the better. A certain degree of ornamentation is allowed on real-estate display maps, but even here, extreme ornamentation will distract attention from the map itself.
5. Avoid heavy black lines or solidly inked spaces, as they give the north point too much prominence. North points shown in Fig. 41 are simple in design, quickly and easily made, and present a neat and dignified appearance.

Border Lines.

To set off the drawing properly, the border line should be one shade wider than the widest line on the drawing. Care should be taken to make the intersection of all corners clear cut, and all border lines the same width.

Single-line borders are to be preferred. If well drawn, they compare favorably with double-line borders, and they take less time.

If double-line borders are used, the lightest line should be on the inside.

Plates 47 to 50 show the types of letters used in topographic drawing. For details, the student is referred to any good text on lettering.

ABCDEFGHIJKLMN
OPQRSTUVWXYZ&

MODERN ROMAN CAPITALS

1234567890

ROMAN NUMERALS

aabcdefghijklmno
pqrstuvwxyz

MODERN ROMAN LOWER CASE

ABCDEFGHIJKLMN
OPQRSTUVWXYZ&

MODERN GOTHIC CAPITALS

1234567890 3

GOTHIC NUMERALS

aabcdefghijklmno
pqrstuvwxyz.

MODERN GOTHIC LOWER CASE

ABDFGIJKLMRSWX

ABGIKLMPRSTWX

abeghmwy — abgmw

2345678 — 2568

ITALICIZED ROMAN & GOTHIC

ABCDEFGHIJKLMNOPQ
RSTUVWXYZ

SHADOW LETTERING

ABCDEFGHIJKLMNOPQRSTUVWXYZ
abcdefghijklmnopqrstuvwxyz 1234567890
ABCDEFGHIJKLMNOPQRSTUVWXYZ
SINGLE STROKE GOTHIC
abcdefghijklmnopqrstuvwxyz 123456789
REINHARDT (SINGLE STROKE)

abcdefghijklmnopqr stuvwxyz 123456789
STUMP LETTERS

ABCDEFGHIJKLMNOPQRSTUVWXYZ
MARGINAL LETTERING

CHAPTER IV

TOPOGRAPHIC DRAWING IN COLORS

Topographic features are often represented by colors alone, or by a combination of colors and the pen-and-ink symbols used in plain topography. This system is little used on engineering maps, since blue prints, photostats, and other common means of reproduction are not adapted to water-color drawings. Color resemblance assists the eye in recognizing various topographic symbols. For example, water is pictured in blue; trees, woods, and vegetation in green; sand in yellow; cultivated land in burnt sienna; etc. Knowledge of these facts enables a layman to decipher much of the information contained on a colored topographic map, which, if mapped in ink, by means of conventional signs, would be entirely beyond his conception. Color topography finds its greatest expression in the landscaping of private estates, public parks, playgrounds, etc., thus enabling the engineer to present the information in the shape of a pictorial map to a private owner or board of laymen untrained in the technique of topography.

Materials and Instruments.

In addition to the instruments and materials required for pen-and-ink topography, the draftsman will need water colors, brushes, blotters, mixing pans, sponge, and washing tumblers, to execute water-color topography properly.

Colors.—Moist water colors in tubes or pans are satisfactory and handy to use. Windsor and Newton's, or other equally good colors should be used, as inferior colors do not mix readily, and soon light fade. The colors necessary are: Indigo I, Prussian blue PB, yellow ochre YO, gamboge G, crimson lake CL, Payne's gray PG, sepia S, burnt sienna BS, to which some draftsmen prefer to add Vandyke brown and Chinese white. The latter is useful in touching up ragged edges and neutralizing strong tints. The colors should be kept in a closed box to keep them moist. Should they become dry and hard, a few drops of glycerin will restore them to good working condition.

Brushes.—At least three sable brushes are necessary, two smaller ones for applying tints, and a larger one for washing and for applying water. Figure 42 shows

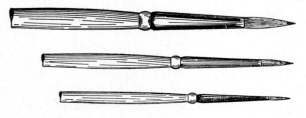

Fig. 42.

the three brushes in full size. The smallest is for tinting very small areas, trimming edges, etc., the second size for general work, and the larger one for toning and rock dragging. To test a brush, proceed as follows:

Wet the new brush thoroughly and shake the water out by a quick downward motion of the hand. The brush should be elastic and show a fine conical point. If the brush shows a tendency to broom or to spread out under this treatment, it is unfit for use. Brushes should always be washed before using, and should never be allowed to remain standing in a glass of water, as this will destroy the flexibility

Fig. 43.

Containers for Color Mixing.—For mixing colors, laying tints, and washing brushes, the draftsman should have the following equipment:

1. A nest of 6 porcelain or earthenware mixing dishes similar to those shown in Fig. 43. These dishes should be 2 to 2½ in. in diameter and not over ½ in. deep.

Fig. 44.

2. Three painted or enamel water pans, about 3 in. in diameter and 2 in. high, as shown in Fig. 44.

3. Two dozen plain blotters about 2½ by 6 in. These will prove indispensable in stopping the flow of water at a given line, removing excess color, etc.

Paper.—Whatman's "cold-pressed" paper is generally used for water-color drawings. It is made in two grades suitable for general color drawing. Grade N has a fine-grained smooth surface, and is in general use for mapping work. Grade R has a coarse-grained rough surface and is used for large-scale drawings and strongly lined work. There is little difference in the two sides of either grade N or grade R, and the general rule is to select the smoothest surface as the working side.

General Rules.

The beginner in color work should first become familiar with the few simple rules given below, whose value he will more and more appreciate as he progresses in the work:

1. Materials and instruments must be kept clean and *free from dust.*
2. *Water* used in preparing tints *must be clean.*
3. The pans of color *must not become daubed* over with other colors.
4. The colors must be *thoroughly mixed and well stirred* at first, and *each time* the brush is refilled.
5. *Use light tints.*

6. A tint must *not* be retouched *in any way* until *dry.*
7. All ink markings must be deferred until the coloring is finished.
8. In preparing a tint from two or more colors, *mix the lightest color first.*
9. Surfaces are best dampened before tinting.
10. The base of the brush must not touch the paper.
11. In touching the sponge to the paper, *don't scrub*, but press *gently* to avoid blemishes from roughness.
12. Repetitions of the *same color* should be of the *same shade.*
13. Shades of *different colors* should have *equal values.*
14. Use care to secure *harmony.*
15. Use *light* ink lines and markings.
16. Work rapidly, but deliberately. Waste no time, but do not get in a panic. Know what you want to do before trying to do it. Have the different steps, their order and manner of execution, well in mind before starting.

Stretching the Paper.

The operation of water coloring being a wet process, it is necessary that the paper remain in a smooth, even surface throughout the process of coloring and drying. To assure this, the paper must be placed on the board in a saturated condition, the edges firmly fastened to the board and allowed to dry in this condition. This operation is called "stretching the paper," and is done as follows:

1. Soak the paper in a flat pan or vessel filled with clean water and sufficiently large to prevent folding

or breaking the paper. Five to eight min. will prove sufficient, depending on the weight and the texture of the paper.

2. Grasp the paper by two adjacent corners and hold it above the pan until all excess water flows out.

3. Place paper on the drawing board in proper position, smoothing out all folds or wrinkles with a small sponge. With a heavy blotter remove the moisture from a band 1 in. wide entirely around the outer edge of the paper.

4. By the use of liquid glue or library paste, cement the edges thus dried to the board. Place thumb tacks at corners and at 6-in. intervals around the border. Allow paper to dry in a horizontal position.

Preparation of Tints.

A *single tint*, i.e., a tint of one color, is prepared by rubbing the wet brush on the surface of the color and stirring it in a quantity of water somewhat larger than the amount needed to color the given surfaces. A *double tint*, i.e., a tint made of two colors, is prepared by first mixing the lighter color with water and then adding the second color until the desired tint is obtained. Tints are sometimes superimposed singly to produce a *double tint*. If properly done, the effect is more striking and brilliant, but should not be attempted by the beginner.

In preparing tints, the colors must be thoroughly mixed with the water by stirring with the brush. Each time the brush is refilled, the entire portion of color should be freely stirred, otherwise a variety of tints may be laid from the same color dish, due to settlement of the color pigment. Water colors are "transparent," and the white of the paper corresponds in effect to the white pigment used in oil colors; therefore, a tint is made strong or light by the use or more or less color, i.e., by varying the amount of water. Light tints should be used by the beginner until he has acquired some skill in estimating color value. If, when a tint dries, he finds it is too light, a second or even a third wash may be given the area until the proper tint is reached. In no case should a wash be repeated until the previous one is entirely dry.

Color Values.

The *primary colors* used are red, yellow, and blue. A *neutral* tint is a mixture of the three primaries given above. *Secondary colors* are produced by mixing two primaries, thus:

Red + yellow = orange
Red + blue = purple
Yellow + blue = green

Tertiary colors are produced by mixing two secondary colors, thus:

    Orange + purple = russet
    Orange + green = citrine
    Purple + green = olive

Blue is a "cold" color, while red and yellow are "warm" colors; therefore, to produce a warm or a cold tint, the predominant primary must be used in greater quantity in the mixture. The character of a secondary tint depends on the character of the dominant primary. Green may be either cold or warm, accordingly as blue or yellow predominates in the mixture. Orange is always a warm tint, since both its primaries are warm colors. Tertiary tints are governed in the same manner.

Where the draftsman is allowed a choice of colors, as in cultivated fields, he should choose colors of adjacent areas to harmonize.

Each of the primaries harmonizers with a combination of the other two, as yellow with purple, red with green, and blue with orange. The same is true of the secondaries. Thus, purple harmonizes with citrine, orange with olive, and green with russet. It must be remembered that the addition of Chinese white does not change the color value of a combination, but simply changes a transparent color to opaque. Unless this is desired, Chinese white should be used sparingly by the beginner.

Laying of Flat Tints.

The laying of a flat, even tint is the most difficult as well as the most frequent operation required of the map draftsman. Clean water, good colors, a soft, flexible brush, and a steady hand represent the minimum essentials for good work. Surfaces 2 sq. in. or less in area may be tinted without previously wetting the surface. All larger areas should be dampened before tinting. To lay a flat tint:

1. The surface to be tinted should be inclined about 5 deg. with the horizontal, sloping toward the draftsman, so that the flow of water in the paper will be in the direction of its greatest dimension.

2. Fill a soft, flexible brush with water and lay a band of water along the upper edge of the area to be tinted. With short rapid strokes of the brush, work the water toward the lower edge, taking care to apply evenly and to keep the right and left edges wet neatly to the penciled line defining the area. When the lower edge is reached, dry the brush on a blotter and pick up excess moisture with the tip of the brush. Retouch any dry spots and trim up the edges. If

the moistened area does not glisten, when viewed obliquely, it is ready to receive the color wash.

3. Prepare the tint as described above and thoroughly stir the mixture. Fill the brush with color and hold in the same position as that occupied by a pen in writing. Start the application of color in the upper left-hand corner of the area, marked *A* in Fig. 45. With a very light pressure, and a motion sufficiently slow to allow complete covering, draw the brush from left to right along the upper edge of the area, then down one-half the width of the stroke, and back from right to left to a position directly under *A*. Repeat this operation with sufficient lapping of the brush to allow free flow of color, until the lower edge of the area is reached. Dry the brush on a blotter and pick up excess color at lower edge by capillary attraction. Work the color to a neat line around all edges. Pick up moisture with the tip of the brush where glistening wet spots appear, and allow to dry. In the initial laying of the color, the pressure should be uniform and the motion continuous from *A* to *B*.

4. A tint *must not be retouched or corrected in any way until dry.* When dry, small light spots may be darkened by stippling (an operation described elsewhere), and small spots darker than the tint may be lightened by alternate applications of small amounts of water from the point of the brush, removing same with a blotter.

5. Large inequalities can be removed only by rewashing the whole area with the large brush before

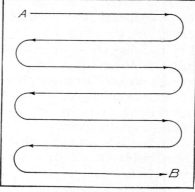

Fig. 45.

the color has had time to set. To do this, use a generous supply of water on the brush and work the color free in the dark spots by brushing. When the area assumes an even tint, take up the excess water by capillary attraction with the large brush.

Precautions.

a. Camel's-hair brushes can be used for blending, picking up moisture by capillary attraction, and moistening paper. They are not elastic enough for laying color. Use only sable brushes.

b. Do not permit the base of the brush to rub the paper when laying a tint, as this will roughen the surface and cause an uneven absorption of color.

c. Use light tints, as you can always darken them with a rewash, if too light.

d. Mix enough color to cover all areas requiring the same tint, as a perfect uniformity in tint can be secured in no other way.

e. In applying color to dry paper, remember that the same color will produce a darker tint than it would if the area had previously been dampened.

f. Avoid dirty pans and brushes, if you wish to lay uniform tints.

g. Do not make heavy erasures on paper previous to tinting.

Laying of Graded Tints.

The preparation of colors and paper surface is exactly the same as for flat tints. To lay a single gradation, first fix the color limits and prepare the tints. Start on the upper edge by laying a band of the lighter color, then adding a quantity of the darker color to the lighter tint. Apply a second band, overlapping the first just enough to form a good union. Repeat the operation, adding the darker color to the tint for each successive band until the extreme is reached. To lay a double gradation, start with the lighter tint and work to the middle of the area exactly as described for a single gradation; then reverse the operation from the middle on by adding water to the heavy tint for each successive band until the lighter extreme is reached.

If the gradation is not satisfactory, allow it to dry and repeat the operation and regrade.

Do not use more color on your brush at any time than is necessary to lay a single band. This is a common fault of the beginner and results in excess water from the band above running into and diluting the color band being laid. This of course results in a streaked and uneven gradation.

Blending of Tints.

To *blend a single tint into white*, proceed to lay the tint as described under Flat Tints to within a short distance of, and parallel to, an imaginary line where the blending is supposed to be complete. Shake the color from the brush and pick up all excess water at the bottom edge of the tint just laid. Wash the brush, fill with clear water, and lay a band of water slightly overlapping the last band of tint. With the point of the brush take up most of the water by passing the brush along its lower edge. Work rapidly and repeat with successive bands of water until the tint disappears into white.

To Blend Tints of Different Colors.

In the rectangle *AABB* (Fig. 46), the upper half is to be tinted green and the lower half burnt sienna, the line of junction to be blended so as to be invisible. Prepare both tints, work rapidly, and cover area *AAaa* with green. Take up most of the superfluous color in rectangle *bbaa*; wash brush and apply burnt sienna on rectangle *bbBB*, working rapidly and stroking away from line *bb*. Remove all superfluous moisture and allow to dry. The beginner will find it convenient to have a brush for each color.

Stippling.

The process of laying a tint by a series of dots placed closed together is called "stippling." The dots should be so spaced that the water flowing from the brush will reach just a little more than halfway to the next dot. This process is used principally in covering spots and blemishes in flat and graded tints and requires considerable skill with the brush.

Stir the color thoroughly and shake the water out of the brush by a quick downward motion of the arm. This operation should bring the tip of the brush to a fine conical point. Take very little color on the point of the brush and touch the point lightly to the paper, with the brush held almost vertical. It is best to cover the space gradually with several sets of dots, always allowing the preceding set to dry before filling in the intervals with another.

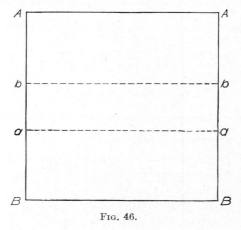

Fig. 46.

Dragging.

Dragging consists in laying a rough, ragged tint used to express rugged surfaces, rock outcrops, shadows, and sometimes, small, irregular water surfaces. The brush is held between the thumb and first finger, making an angle of about 5 deg. with the paper, as illustrated in Fig. 47. The hand is supported by the tip of the third finger resting lightly on the paper and the marking is done with the side of the brush. The strokes are parallel to the contours in rock dragging and parallel to the bottom

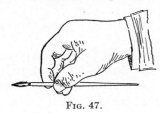

Fig. 47.

border in swamp dragging. The stroking is continued until the proper tint is secured. This operation cannot be hurried, and care must be taken not to smear or smudge one stroke into another. A second and third dragging is often necessary to secure the proper degree of ruggedness, but one application must be thoroughly dry before a second is attempted.

Precautions.

1. Mix colors thicker than for laying tints.
2. Carry very little color on the brush.
3. Always try the brush to be sure it will not smear, before using.
4. Never attempt to drag a surface that is not completely dry.
5. Do not attempt to erase or wash out dragging, but retouch with a smaller brush.

Conventional Signs and Tints.

Crop symbols, fences, and property lines are drawn in ink, as in plain topographic drawing. The inking must be deferred until all coloring has been applied, and allowed to dry. Cultivated crops, such as tobacco, corn, cotton, etc., have a background tint of burnt sienna with the crop symbol superimposed in india ink. Fruit crops, orchards, vineyards, etc., have a background tint of warm green, with the ink symbol superimposed.

Cultivated Land.—The conventional tint for cultivated land is a light shade of burnt sienna, ruled over with a right-line pen in a slightly darker shade of the same color. The ruling is done with T square and triangle similarly to section lines in mechanical drawing. The weight and spacing of lines is varied to suit the scale and texture of the drawing. On drawings showing much detail, the monotony of color is often relieved by tinting alternate fields crimson lake and

Payne's gray, ruled with a darker shade of the same color. If cross-hatching is omitted, as it often is on small-scale maps, the only conventional sign permitted for cultivated land is burnt sienna.

Cleared Land.—A warm, flat tint of Prussian blue and gamboge is used to represent this symbol. Some draftsmen use a cold tint, *i.e.*, one with the blue predominating, for cleared land, and a warm tint with the gamboge slightly dominant for meadow and pasture land.

Underbrush.—The conventional tint for underbrush is a combination of the cultivated-land and cleared-land symbols laid in alternate splotches to produce a mottled-green and burnt-sienna tint, and is executed as follows:

1. Moisten area to be tinted as in laying flat tints.
2. Prepare the separate tints as for cleared land and cultivated land.
3. Remove excess moisture until paper is not quite so wet as for laying flat tints.
4. Use two brushes and stipple alternate splotches of the above colors.
5. Take a small pointed brush and blend the colors where they come in contact.
6. Shake water out of large brush and remove excess color with the point of the brush.
7. Lay in a horizontal position to dry.

Marsh.—Prepare and lay the tint for underbrush as a background for the marsh symbol. Drag with Prussian blue as per instructions under Dragging. Care must be exercised not to make the dragging too pronounced. Some draftsmen prefer indigo for the dragging operation, because it is more subdued in color.

Sand.—The conventional sign for sand is a flat tint of yellow ochre.

Gravel is represented by superimposing a series of dots made with Gillott's 170 and 303 pen upon a background of the sand symbol. The color used for the dots is burnt sienna. *Coarse gravel*, or gravel and stone mixed, is shown by scattering a few rounded and triangular shapes throughout the gravel symbol.

Mud.—Use a very light tint of sepia as a background wash, then superimpose the ruled lines of a slightly darker shade of sepia upon the flat tint, as in cultivated land. The same rule for breaking of lines to represent sun cracks should be observed as in pen topography.

Water.—Use a flat tint of Prussian blue, the shore lines being represented by the same color drawn with Gillott's 303 pen. If a lower tone is required, use indigo instead of Prussian blue. On display maps, water is often represented by a graded tint of Prussian blue, laid with the darkest band of color at the shore line and blending into white as it approaches the center of the stream. For lakes, bays, and ocean shore line, the graded band of shore-line color blends into a very light flat tint of the same color.

Trees.—Shrubbery, trees, and clumps of trees are first outlined in pencil as a guide to the formation of high lights and shadows.

On a large-scale map, a flat tint of gamboge is used over the entire surface outlined by the penciled areas. The shading is then accomplished by applying a darker and warmer tint than the one used for cleared land. If the water features are in Prussian blue, this tint should be composed of Prussian blue plus gamboge. If indigo, the tint should be of indigo plus gamboge. The shading strokes must parallel the outlines of the tree, to bring out its rounded form. When the proper degree of shading has been reached, allow to dry, and touch up the previously penciled outlines with the same color, applied with a freehand pen. The addition of a small amount of sepia will improve this tint, especially if Prussian blue is one of its component colors. Some draftsmen prefer to stipple the shaded strokes, and the effect is very pleasing, if well done.

Oak Trees.—Since oak trees are drawn much more irregular in outline than deciduous trees, the shading can be best accomplished by stippling. A warmer tint of the same colors as used for deciduous trees should be applied to oaks. This provides a color distinction which will often compensate for a lack of angularity of outline.

Evergreens.—Use a somewhat colder tint of Prussian blue and gamboge. Outline as for other tree forms and apply shading. The outlines are then covered by a series of radial hachured strokes representing spines. The spines are closely spaced on the shaded side of the tree and sparsely or entirely absent opposite the high light. Draw the spines sharp and distinct, but not long enough to mar the appearance of the tree.

Orchards.—Use the symbol for deciduous trees, symmetrically spaced on the background tint representing cleared land. A shadow is added to the tree symbol as in pen topography and is composed of a light tint of Payne's gray and crimson lake. The shadows should conform in direction, size, and shape, to the rules given under Orchards, in Pen Topography. A neat, shadowy effect may be secured by painting in the shadow and almost immediately taking up the excess color with a blotter.

Woods.—For woods and trees on small-scale maps, use a flat tint of Prussian blue plus gamboge warmed with sepia. Omit shadows, shading strokes, and high lights, but define tree clumps by outline strokes.

Buildings.—The conventional tints for buildings are sepia for frame, crimson lake for brick, and Payne's gray for stone. Use a flat tint for the building proper, and outline with ruled lines of a very strong tint of the same color. The only exception to this practice is that frame buildings are often outlined in india ink.

Roads, Trails, and Paths.—On large-scale maps, fill the penciled outlines with a flat tint of yellow ochre. Allow to dry, and draw the outlines in india ink, using the same symbols as for pen topography. On small-scale maps, the flat tint is usually omitted.

Bridges.—The outline of the road crossing the bridge is filled with a flat tint of yellow ochre, and the same distinctions as to wood and masonry are observed as in buildings.

Shading of Slopes.—Cuts and fills trimmed to an even slope should be represented by a graded tint of burnt sienna plus sepia, with the heaviest shade of color applied to the highest part of the slope. Irregular and eroded slopes are best shown by hachures. Use a limber pen and the above color applied as ink.

Rock Slopes.—The conventional color is sepia warmed with burnt sienna. Apply by dragging over the previously shaded slope to produce the general forms of rock outcrop. Keep strokes parallel to the contour lines and repeat a second or third time, if necessary. Bring out the lights, shadows, and irregularity of outline by a few touches of strong color, made with an ordinary pen.

Shading of Slopes.—Use graded tints of Payne's gray and crimson lake, enough of the first ingredient to produce a bluish tinge.

1. Prepare several gradations of the above tints. Usually four or five are enough.

2. Draw the contours in fine distinct pencil lines to use as a guide to intensity of color.

3. Lay a flat tint of the lightest shade over the entire area to be shaded. Blend the edges into adjoining colors, and allow to dry.

4. Slightly moisten the areas to receive a darker shade and apply the various shades of colors according to the contour spacing.

5. Carry very little color on the brush at one time and either stroke parallel to the contours or apply by stippling.

6. Allow the drawing to become thoroughly dry when light spots may be corrected by stippling and dark spots lightened by washing and blotting.

Contours.—Contours are drawn in fine distinct lines of crimson lake after the drawing is thoroughly dry. Mix the color on a wet brush and transfer to the freehand pen. Ink with very little pressure, always drawing the pen towards the draftsman. Do not hurry this operation, as the roughness of cold-pressed paper makes inking difficult.

India Ink.

Lines drawn in india ink have a general tendency to overshadow the details of a water color drawing, and are therefore made much lighter than in pen topography. Three grades, or widths, are used. Property lines, fences, outlines of cultivation, and contour lines are made as fine as they can be clearly represented.

Secondary roads, trails, paths, shore lines, and outlines of buildings are made slightly heavier. The heaviest lines are reserved for important highways, railroads, and bridges. For weight of lines not mentioned above, follow the general rule that the breadth of lines should be proportioned to the relative importance of the objects they represent.

Suggestions for Laying Tints.

1. Details must be accurately outlined in pencil before tinting begins.

2. With an art-gum eraser remove the heaviest of the penciling, leaving just enough to use as a guide in coloring.

3. Tints must be of such relative intensities as to make them readily distinguished from all others. Small surface areas should therefore be slightly darker than larger areas of the same tint.

4. The green for trees and cleared land should be a cold tint, with blue predominant.

5. Proper gradation and relative color values are easier to obtain if the lightest tints are laid first.

6. All dust should be removed from the drawing before tinting begins. Wash with a small sponge and running water, and allow to dry.

7. Erasures, if made at all, must be very lightly rubbed with art gum. Never use a hard rubber eraser.

Durability of Colors.

All water colors undergo more or less change from exposure to light and atmospheric impurities, the cheaper colors notably so.

Of the colors used in mapping, Prussian blue and crimson lake are the most sensitive and unreliable. Indigo, Payne's gray, and sepia are affected to a much less degree; while burnt sienna, yellow ochre, and gamboge may be regarded as permanent, except for the fact that all water colors become darker from exposure. A water-color map, to stand the maximum of exposure, should have cobalt blue substituted for Prussian blue and light red for crimson lake. Tints of indigo, sepia, and Payne's gray should be applied with greater intensity than would otherwise be required.

Blocking Out and Penciling Details.

The plotting of traverses, contours, trees, and other details must be done with extreme care to prevent soiling or injuring the surface of the paper. Pencil lines must be clear, yet faint enough not to interfere with the laying of colors. The pencil should be sharp and hard enough to give a faint line, yet not sufficiently hard to injure the surface of the paper. It is best to keep the paper partially covered with a paper shield during the operation of plotting. In the immediate portion, where work is being done and where the plotting involves the use of many lines and numerous details, it is advisable to plot the details on a separate sheet and transfer the same to the stretched paper. This prevents soiling the paper and making subsequent erasures. The space for title, north point, and legend should be blocked out at this time, and protected by a shield, until the map is finished.

Lettering.

The intense black of india ink, when placed in juxtaposition to colors, will have one of two effects. It will either give the impression of overemphasizing the lettering, or the counter effect of subduing the colors. Since neither effect is to be desired, the lettering is made as light and as inconspicuous as possible. Where single-stroke lettering is used, the strokes are light and slender, and modern Roman is inked in outline only. When the lettering is commercial Gothic, the letters are of the shadow type, as shown on Plate 49. To get this effect, only those parts of the outline lying on the shaded side of

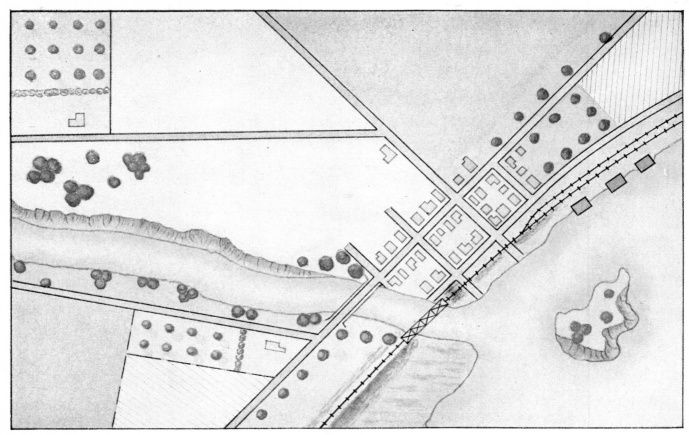

PLATE 53.

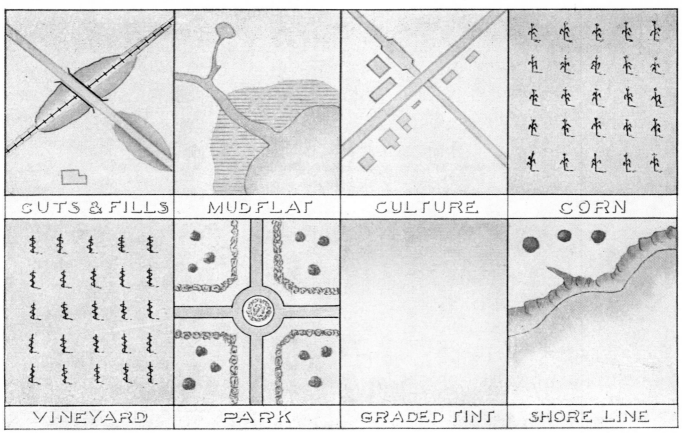

PLATE 52.

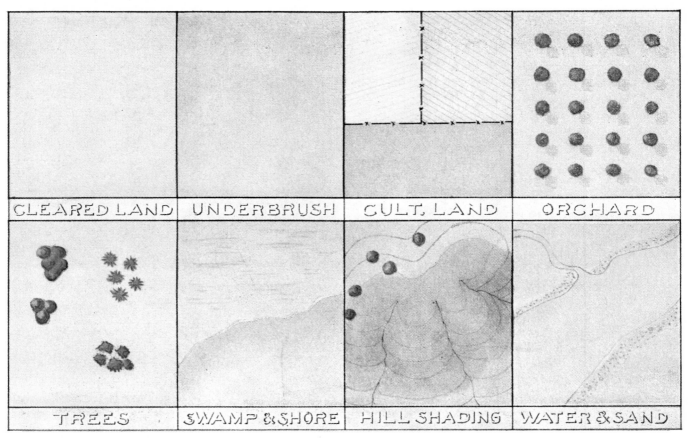

PLATE 51.

the stroke are inked. The light is considered as coming from the upper left hand corner at an angle of 45 deg. On rounded letters, such as O, C, G, and Q, the shaded line starts at a point where a 45-deg. line becomes tangent to the curved stroke of the letter.

Plates 51, 52, and 53 show the various topographic symbols, reproduced in color.

The Use of Colored Pencils.

The draftsman inexperienced in the use of water colors may make a very presentable colored map by the skillful use of colored pencils.

These pencils are manufactured in all the standard colors used on water-color maps and are available in both light and dark shades of the same color. The pastel shades of green, burnt sienna, blue, and yellow ochre are used to produce the delicate background tints required for colored maps. The darker shades of green are used for orchards, trees, hedges, and crop symbols. Payne's gray is used for hill shading, burnt sienna and sepia for rock dragging, and indigo or Prussian blue for swamp dragging and shore-line shading. Many draftsmen prefer to lay the background tints in water color and use colored pencils to superimpose such symbols as trees, orchards, houses, etc., after the background tints are thoroughly dry. Mistakes made in pencil are easily corrected, and much time may be saved by using pencils and water colors in combination.

Maps made by the use of colored pencils or pastel chalk smear or smudge easily and are not permanent. They may be made fairly permanent by being sprayed lightly with a fixing solution of very thin varnish or clear shellac. Care must be exercised not to spray the drawing too generously as this treatment would give the map a glossy surface and ruin its appearance.

Colors Used on Geologic Maps.

The standard colors for the 12 systems of sedimentary rocks as shown on the U. S. Geological Survey folios are as follows:

1. Quarternary.....................Ochraceous orange.
2. Tertiary........................Yellow ochre.
3. Cretaceous.....................Olive green.
4. Jurassic.......................Blue-green.
5. Triassic.......................Light peacock blue.
6. Carboniferous..................Blue.
7. Devonian......................Gray-purple.
8. Silurian......................Purple.
9. Ordovician....................Red-purple.
10. Cambrian.....................Brick red.
11. Algonkian....................Terra cotta.
12. Archean......................Gray-brown.

If there are two or more formations in any one of the above systems they may be distinguished by different patterns of straight parallel lines of the same color. Patterns for subaerial deposits are composed of dots and circles and may be printed in any color, although yellow is most frequently used.

Colors for igneous rocks are more brilliant than those used for the sedimentary series. No particular colors are prescribed, but pink or red is to be preferred.

Small areas are colored solid while some cross-line pattern is used for larger areas.

Metamorphic rocks are designated by short dashes irregularly spaced. These dashes may be in black ink or in color over a ground tint of lighter shade. They are sometimes shown as white lines on a ground tint or pattern. Since practice is not uniform as to colors used in the United States and other countries, each geological map is usually accompanied by an index and a color legend.

CHAPTER V

CONTOURS AND CONTOUR SKETCHING

To become expert in topographic drawing, it is necessary not only to become skilled in lettering and freehand drafting, but also to acquire a certain degree of proficiency in portraying, by means of contours, widely different forms of relief found in nature.

A knowledge of geology sufficient to recognize and portray such contour forms as are common to volcanic, glacial, and eroded regions is an invaluable aid, both to the topographer and the draftsman. By a study of contour maps showing regions of different geological origin, the student can become familiar with the above forms of expression and even develop certain mechanical skill in delineation; yet, without a clear understanding of the contour, its use in representing slopes, ravines, ridges, depressions, etc., he will fail to portray accurately the topography of a given region. The draftsman with a limited knowledge of contour forms too often confines his interpretation to forms with which he is familiar, although they may utterly fail to portray the topography he is mapping. Thus it may happen that two draftsmen working on adjacent sheets of similar topography will draw maps very unlike in appearance. The one having received his previous experience in rough, rugged country will be likely to introduce too much angularity into his work, giving it the appearance of fine-textured topography, while it actually should be represented as coarse textured. The other, having received his training in gently rolling country, will be likely to introduce long, sweeping curves into his contour work, making the map meaningless, and failing to show the proper drainage. It becomes apparent that in order faithfully to portray relief, the draftsman must have an intimate knowledge of contours and contour forms.

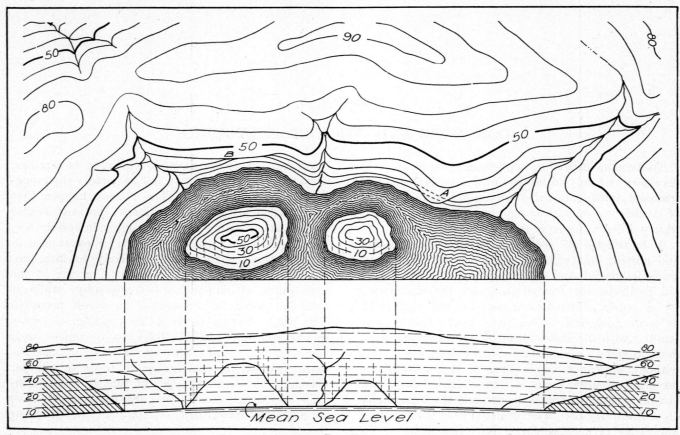

Fig. 48.

Figure 48 shows a water surface used as a datum and a series of level surfaces, spaced 10 ft. apart vertically, cutting out a series of contours pictured in the adjoining sketch.

Contours.

The intersection of the surface of the ground by a level surface is called a "contour," and the projection of this intersection on the map is called a "contour line." The vertical distance between two level surfaces cutting out adjacent contours is known as the "contour interval," while the level surface, to which all contours are referred, is called the "datum." For example: If mean sea level is taken as the "datum" or surface of reference, the shore line would represent a contour of zero elevation. If the tide should rise 10 ft., the new shore line thus formed would be a contour whose elevation is 10 ft. above datum. If the water should continue to rise until the land is submerged, the successive shore lines formed at each 10-ft. rise would give us a contour map of the land whose contour interval is 10 ft.

By referring to the above sketch, certain facts concerning contours become apparent:

1. A contour is a line all points of which lie in a level surface.

2. Every contour closes upon itself either within or without the limits of the map; in the latter case it is drawn to the edge of the map.

3. A contour line closing within the limits of the map either indicates a summit or a depression. If the depression does not enclose a body of water, it should be numbered or hachured to indicate its elevation.

4. Contours never cross each other, except in the case of an overhanging cliff or a cave, and then they must cross twice. Such cases seldom occur, but if they do, the lower contour should be dotted (see A, Fig. 48).

5. On uniform slopes, contours are evenly spaced.

6. On a plane surface, contours are straight lines, parallel to each other.

7. The distance between contours varies inversely as the slope.

8. The sharpest bends in contours occur at their intersection with ridge and valley lines, which they cross at right angles.

9. Contours should not be drawn across streams. As they approach the stream, they turn upstream and disappear, in coincidence with the shore line, to reappear directly opposite on the other shore line. This rule is often violated on small-scale maps such as the U. S. Geological Survey quadrangle sheets.

Here the contour is carried directly across the stream at right angles to the shore line, to avoid the loss of identity of the contour.

10. A single contour cannot intervene between two other contours having the same elevation either on a summit or in a valley, *i.e.*, the maximum ridge and minimum valley contours must occur in pairs if they do not close within the limits of the map.

11. One contour cannot be superimposed upon another except where they indicate a vertical cliff (see *B*, Fig. 48).

12. Contours bend toward the upgrade when crossing a valley or depression, and toward the down grade when crossing a ridge line.

13. Contours crossing a railroad laid to an even grade will be spaced at equal intervals.

Contour drawings are not a complete representation of natural relief, although they do offer an abstract picture of the terrain. To compare a contour drawing with nature, in the same manner as other pictorial representations would be compared, is impossible, because the earth's surface can never be entirely seen from any one position to the extent in which it is there reproduced. This is also the reason why contour maps, navigation charts, landscape drawings, etc., are just so much unintelligible hieroglyphics to those who are not familiar with topographic representation. Present the same map to a trained geologist or topographer and he will not only gain a complete mental picture of the relief, but will also obtain much other valuable information, such as: character of rock formation; presence or absence of coal outcrops; character of soil, as shown by erosion; etc. Some experts may even go so far as to name the part or parts of the country where the map was made. A map so contoured as to show at a glance the type of country mapped, as flat, hilly, rolling, or badly eroded, is said to have "topographic expression."

Limits of Topographic Expression.

The faithfulness with which topography may be shown by contouring is limited by the scale of the map, the contour interval, and the experience and skill of the topographer.

If the scale of the map should be reduced on a badly eroded section of country, until the reentrant contours showing gullies and defining drainage become mere indentations in roughly curved lines, the map would fail to express a true conception of the relief. Likewise, if a small contour interval were chosen to delineate a rugged, hilly country, the contours would merge together on the steeper slopes and obscure the most prominent features of relief.

Granting the skill and experience of the draftsman, the choice of scale and contour interval place definite limits on his ability to faithfully express by means of contouring the true relief of the region.

Contour Sketching.

This term is used advisedly, because contours are never completely surveyed as are features of culture, but have a control of salient points whose elevation and location is fixed by field measurement. The relief is then pictured to the scale of the map, by contours drawn freehand with their position controlled at convenient points and sketched between these points. However numerous these control points may be, they are always at an appreciable distance apart, and it is possible to interpret the contours so as to give more than one shade of meaning to the map, although both interpretations may be within the authority of the control points. It is therefore significant that the map giving an interpretation nearest the truth will be drawn by the draftsman having the fullest knowledge of the origin and character of the relief feature expressed.

In small-scale topography, it is especially desirable that the draftsman have at least a working knowledge of the physiographic processes shaping the earth's surface, and be able to recognize and execute the peculiar contour form representing the same. A full discussion of the relations existing between physiography and topographic drawing will not be attempted in this text, but rather an attempt made to point out and illustrate a few of the typical contour forms

Fig. 49.—Glaciation—mountain.

which serve as a key to the physiographic development of a given region. For a complete study of glaciation, erosion, upheavals, winds, and other natural forces engaged in changing the outlines of the earth's surface, together with the contour forms peculiar to each, the student is referred to various texts on physical geography and topographic geology.

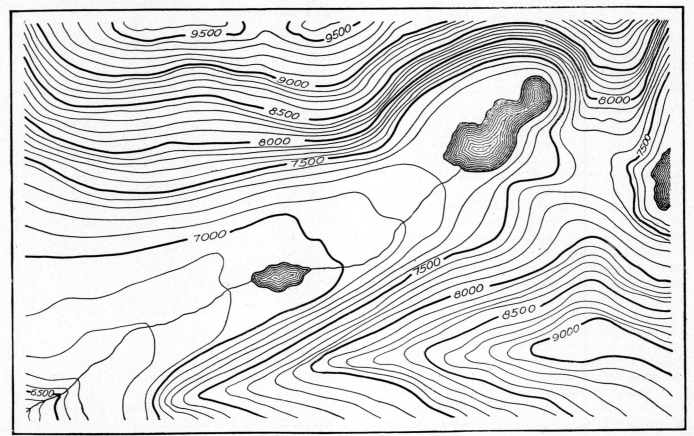

Fig. 49a.—Glaciated valley.

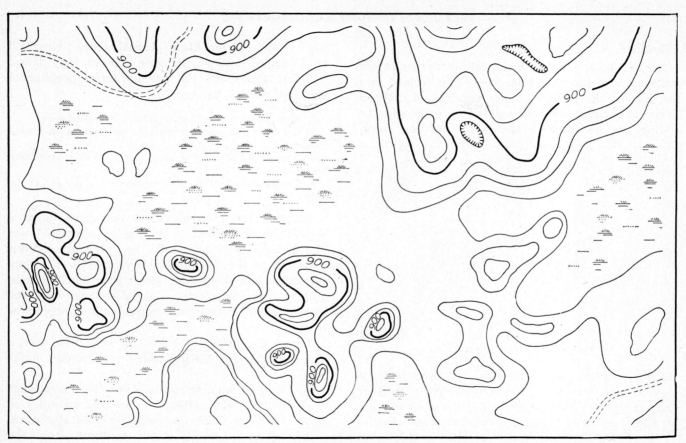

Fig. 50.—Glaciated area—flat country.

Glaciation.

Large masses of snow accumulate in the mountain valleys above the snow line, where it is formed into ice, due to the pressure of its own weight. When sufficient weight has accumulated to overcome the sliding friction between the earth and the snow,

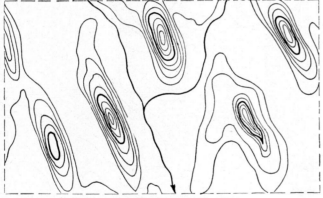

FIG. 51.—Glaciation—drumlins.

the mass begins to move down the valley, gathering a load of stone, gravel, and soil in its progress. Smaller glaciers unite to form larger ones in the same manner in which a river is formed by its tributaries. Figure 49 illustrates the formation of glaciers in the radial valleys of a mountain peak. The contour forms showing evidence of mountain glaciation are "cirques," or U-shaped valleys, hanging valleys, and lakes.

Figure 49a shows the contour forms peculiar to glaciated valleys, namely:

1. Straight, even slopes terminating in a U-shaped end.
2. Contours not pointing sharply upstream as in eroded valleys.
3. Absence or scarcity of small, side gullies or tributary streams.
4. Tendency to form small lakes near the upper end of the valley.

Hanging valleys result where one glacier joins a second whose bottom is flowing at a much lower elevation. They have all the peculiar contour markings of the U-shaped valleys, except that they terminate in a steep slope where they join the main valley. Lakes result in glaciated valleys when the glacier meets a hard rock formation and must flow over it.

Figure 50 shows the effects of glaciation over a flat terrain. The distinguishing topographic markings are:

1. The long, winding contours have been somewhat changed by glacial action, but are still an index or guide to the general topography before glaciation.
2. The many small, closed contours show the depositions from a lateral moraine.

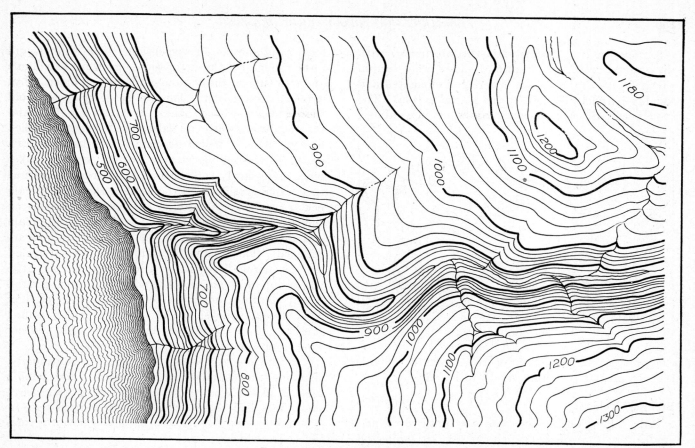

Fig. 52.—Eroded valley—first stage.

3. The resulting swamps show the interference with the original drainage. Figure 51 shows a flat terrain covered with drumlins, *i.e.*, a series of elongated and roughly parallel hills left by the glacial action. The same peculiarities of contour may be noted as above,

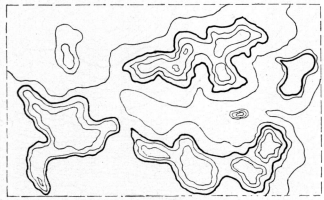

Fig. 53.—Erosion cycles.

except that the elevation of the separate mounds is much greater than that shown in Fig. 50.

Erosion Cycles.

Rain falling on a newly formed continent would erode it in the following manner: Water falling on the slopes would seek a lower level, carrying with it loose sand and soil in suspension. The water would form rivulets in the depressions, growing in volume and velocity as the various streams combined, until it reached the ocean. The first stage or cycle would show steeply eroded banks, rapid stream fall, and practically no width of valley except that occupied by the stream. When the main stream bed reaches a hard, wear-resisting strata, bottom erosion takes place more slowly and the tributary streams continue their rapid erosion until the level of the main stream bed is reached. As erosion on this level recedes up the tributary streams, the valley of the main stream and the lower stretches of the tributary valleys are widened. This might be termed a "second-cycle" or "second-age" period. In a section having little vegetation and an easily eroded soil, this process would continue until the terrain was composed of wide stream valleys, low divides, and gently rolling country. This condition represents still another cycle of erosion.

The process of erosion thus goes on continually, changing the character of the uplands and joining coastal plains and wide valleys along the ocean shore lines and rivers. Each of the various ages or cycles of erosion has its own peculiar contour formations, by which it may be recognized. Figure 52 shows a stream comparatively young, the distinguishing marks being steeply eroded banks, narrow valleys, and contours pointing sharply upstream and having few tributaries. Compare this with rivers having wide valleys, and note the different contour forms. Figure 53 represents a section having been eroded into a very gently rolling terrain, except for certain parts, shown by closed contours, which were formed of a much harder strata than that immediately surrounding them. The long, continuous contours show the general stage of the erosion and define the drainage, while the closed contours define the limits of the harder, slow-eroding strata. This form might be confused with glaciation of a flat terrain but for the absence of swamps and impeded drainage, and the angularity of the contours as compared with drumlins.

Alluvial fans may be formed either at the mouths of eroded canyons in block-mountain formation, or by tributary streams in the process of river-valley widening. Figure 54 shows alluvial fans at the base of a mountain formation rising directly out of the plain, while the sketch in the right-hand corner shows the start of alluvial fans along a stream. The soil, gravel, and stone carried by the stream is deposited by the water as it spreads out fan-shape at the mouth of the canyon. If the streams are close together, these fanlike deposits will merge, and, when shown by contours, would appear as a series of flat loops with their identations pointing towards the higher ground. The contours are often shown as straight parallel lines, designating a gentle, even slope, by topographers not familiar with their method of formation. Alluvial fans may also be formed in conjunction with triangular facets occurring along a fault, *i.e.*, an abrupt, almost perpendicular rise, occurring along a straight line and caused by an earthquake or an upheaval. The erosion of gullies in the higher ground results in sharp, angular contours where the sides of the valley intersect the face of the fault. The entire character of the formation may be misrepresented by using easy, sweeping curves to connect a contour on the side of the valley with the same contour on the face of the fault. This is a common error with many topographic draftsmen. The contours representing the alluvial fans are similar to the ones shown in Fig. 54, except that they are more regular and even.

Wind Erosion.

Many curious and varied forms are sculptured by the combined action of sand and wind and have no particular contour index. Only one example is shown in Fig. 55 because of its similarity to glaciation in flat terrain. As in Fig. 50, the contours extending

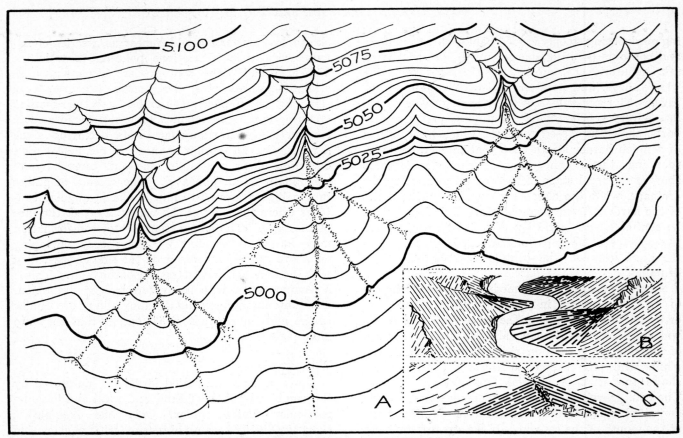

Fig. 54.—Stages in the formation of alluvial fans.

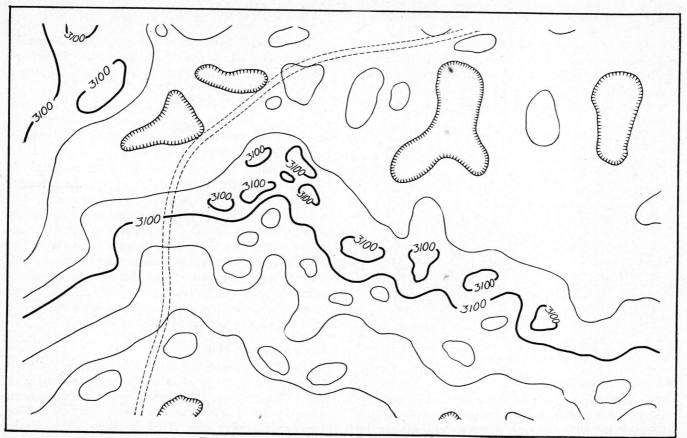

Fig. 55.—Contours and sand country sculptured by wind.

entirely across the map show the general topography, while the numerous closed contours indicate small sand dunes or depressions formed by wind action. Wind action may be distinguished from glaciation by:

1. Closed contours, indicating lower and smaller hills than in glaciation.
2. Absence of swamps and impeded drainage.
3. Absence of water inside of depressed contours.

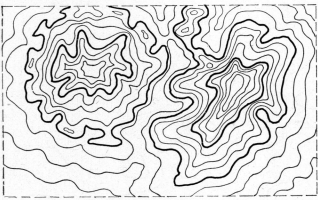

Fig. 56.—Volcanism.

Mountain Formations.

The folded type of mountain formation may be readily distinguished by the roughly parallel ridges separated by wide, smooth valleys, and cannot be confused with other formations.

The general form of dome mountains is easily distinguished from the block type, but is not so readily recognized when compared to peaks having a volcanic origin.

The marks of volcanism shown in Fig. 56 are:

1. Radial valleys and ridges.
2. Ridges increasing in height as they approach the summit.
3. Marks of ash cones or craters.

The distinguishing form of the dome mountain is the encircling row of hog-backs. If the dome mountain has radial valleys and is badly eroded, the hog-backs may be partially concealed by the formation of alluvial fans against the steeper slopes of the hog-backs which face toward the summit. In the above case, only the correct mapping of the hog-backs will differentiate between these forms. Block-mountain formation results from geological faults, and can be recognized by triangular facets which occur where eroded valleys intersect the face of the fault.

Use of Contour Maps.

The U. S. Geological Survey in cooperation with the various states is making a topographic atlas of the United States. This work has been in progress since 1882 and is about 47 per cent complete. This survey has been completed in Ohio, New York, Connecticut, Delaware, Maryland, Rhode Island,

Massachusetts, New Jersey, West Virginia, New Hampshire and the District of Columbia, and is being as vigorously prosecuted in the remaining states as federal and state appropriations will permit. The atlas is being published in sheets 16 by 20½ in., covering 15 min. of latitude by 15 min. of longitude. The area covered in each sheet is called a quadrangle, and each quadrangle is given the name of a principal city or some prominent natural feature within it. The scales are 1:31,250, 1:62,500, 1:125,000 and 1:250,000. These scales are fractional parts of 1:1,000,000, which has been adopted as the scale for the final production of a world map.

Quadrangle sheets may be purchased in lots of less than 50 from the director, U. S. Geological Survey, Washington, D. C.

As an aid in the study of U. S. Geological Survey quadrangle sheets, the authors have added to this chapter a Glossary of Topographic Forms.

The student should become familiar with the topography that the terms defined in the glossary represent and learn to express these terms with contour sketches.

It is only by a close study of the various topographic forms that the draftsman is able to represent by contours a true picture of the drainage and terrain.

As an illustration, several terms defined in the glossary are represented by small perspective drawings, others are shown both in perspective and by contours, and a few are represented by contours in plan view and by a cross-section view showing the profile. The knowledge gained by this method of study will add to the student's ability correctly to interpret quadrangle sheets showing various forms of topography.

These quadrangle sheets furnish a vast field of information valuable both to the draftsman and the engineer, and from them many preliminary studies may be made which otherwise would call for extensive field surveys. For example:

1. The drainage area of any watershed complete within the map may be determined by planimeter.

2. The impounded area and approximate cubical contents of a proposed reservoir may be found in a similar manner.

3. The profile and alighment of a proposed route may be taken direct from the map and approximately determined by an office study.

4. The latitude and longitude of any point on the map may be determined by scaling sufficiently accurate for calculating true north in ordinary transit surveys.

5. The nearest bench mark to any particular job may be located from the quadrangle sheets.

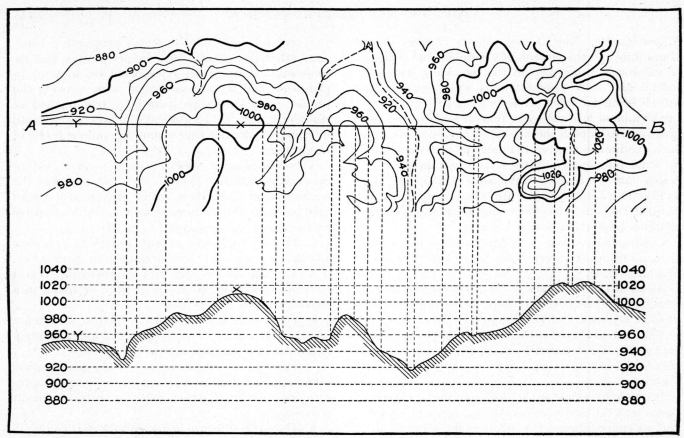

Fig. 57.—Construction of profile from plan.

6. High- and low-water marks are given at intervals along all important streams.

Construction of Profile from Plan.

To construct the profile of any given line as AB (Fig. 57), proceed as follows:

1. Pin the map containing the line AB, the profile of which is desired, firmly to the drawing board.
2. Draw horizontal lines spaced in vertical scale to suit the contour interval. These lines should contain the elevation of both the highest and lowest contour crossed by the line AB.
3. Orient the paper on which the profile is to be drawn until these lines are parallel to AB.
4. Project all intersections of line AB and the various contours crossed to the line on the profile representing the contour elevation. This gives a series of points all lying on the profile.
5. Intermediate points, such as y, are interpolated by scaling the proportionate distances the line AB lies from the 940 and 960 contours.
6. Points such as x cannot be properly interpolated unless the location and elevation of the highest point lying inside the 1,000-ft. contour are known.

Site Plan.

In planning and designing bridges, dams, factories, etc., the engineer and the architect often require large-scale maps showing the site of the proposed structure. Such maps are called "site plans," and contain much detailed engineering information. Figure 58 is a site plan for a proposed bridge, with its attendant profile of center line. Information included on this map, but not usually shown on a topographic map, are high-water marks and subsoil, or foundation, conditions.

Drainage or Flooded Area.

The heavy dotted line in Fig. 59 is a line bounding the drainage area above the dam site shown on the extreme right. Such a line is located by starting at crest elevation on the dam and ascending the line of steepest declivity until a summit between two streams is reached as at a. The line should then follow the summit between streams until it has traveled completely around the headwaters of the stream above the dam site. When this point, marked b on Fig. 59, is reached, descend in the line of steepest slope to the crest of dam. This irregular area is measured by planimeter, and run-off is calculated from rainfall data.

To determine a flooded area of a proposed damsite:
1. Trace the contour corresponding to the crest elevation of the dam up the valley until it crosses the

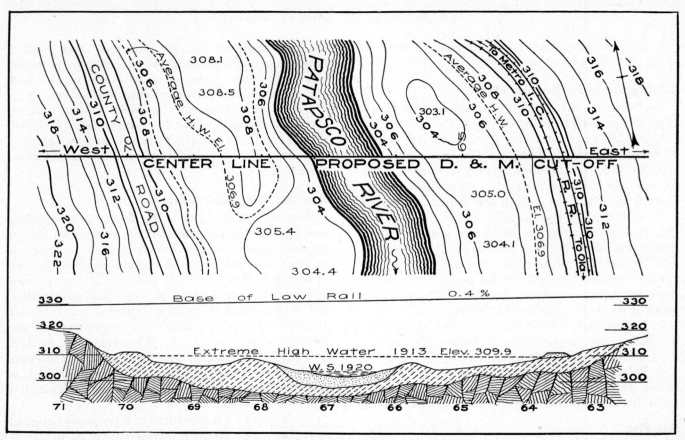

Fig. 58.—Site plan.

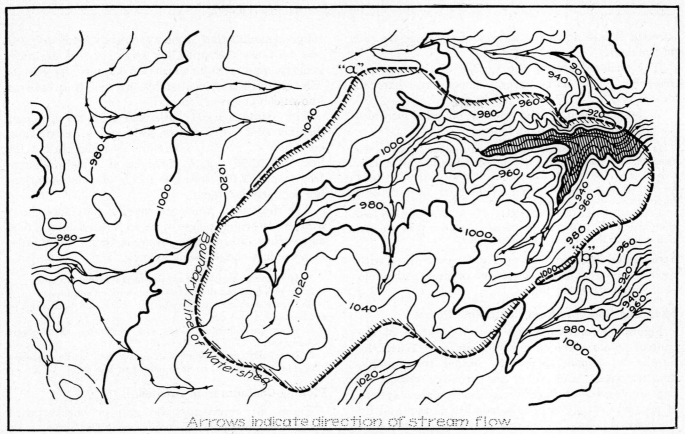

Fig. 59.—Drainage area, flooded area, and dam.

stream, thence down the stream to the opposite end of the dam.

The enclosed area is the surface of the proposed reservoir. The above area is shown cross-hatched on Fig. 59.

A rough estimate of the water impounded may also be taken from the map, as follows

1. Measure the area enclosed by each contour shown on the reservoir site, including the one corresponding to crest of dam.
2. Average the areas of the adjacent contours and multiply by the contour interval. This will give the cubic contents impounded between contours.
3. When all the horizontal sheets of water, each of one-contour interval in depth, have been calculated, their summation will give the capacity of the reservoir.

Stream and drainage lines on Fig. 59 are marked with arrows to indicate the direction of flow and avoid confusing with other lines.

Sketching Contours from Controlling Points.

Figure 60 represents the plotted position of the control points taken in a stadia survey. When the number of control points is limited to summits, divides, stream and gulley lines, road intersections, and profiles, the draftsman must interpolate a system of contours satisfying all the conditions required by the control points. This will at best give him but a rough generalization of the topography being mapped, and he must call upon his knowledge of drainage, geology, and contour forms to supply the details. The contours are located by interpolating between known elevations as follows:

The points marked 1,103 and 1,124 have a map distance of 29 divisions of scale and a difference of elevation of 21 ft. Therefore, the elevation 1,120 lies $17/21 \times 29 = 23.4$ divisions of scale measured from point 1,103 towards 1,124. Other elevations are similarly located.

The draftsman must exercise his judgment, however, as to which points were so located in the field as to allow interpolation between them. The order of procedure should be as follows:

1. Interpolate position of all contours crossing streams.
2. Interpolate and draw closed contours on summits.
3. Locate contour crossings in depressions on both sides of saddle as at A and B.
4. Draw in indicated roads and determine all contours crossing them.
5. Interpolate remaining points and sketch in the contours.
6. The result should be as indicated in Fig. 60.

Plotting Contours from Profiles.

Figure 61 represents the grid or checkerboard method of interpolating contours. The area to be

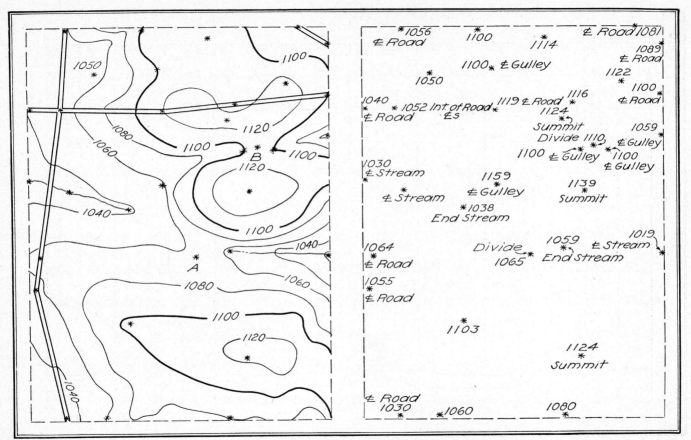

FIG. 60.—Contours plotted from control points.

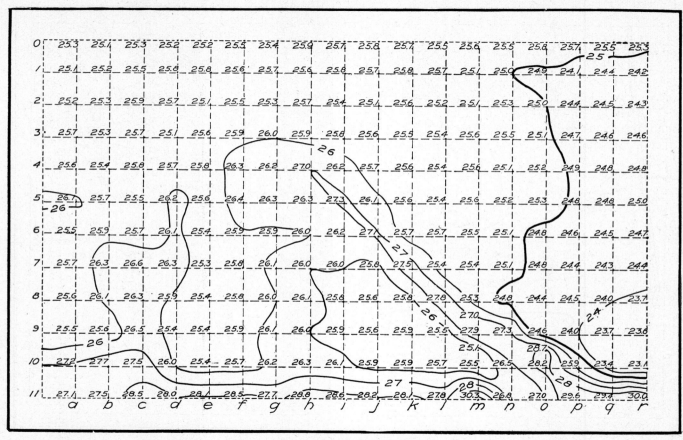

Fig. 61.—Checkerboard map.

contoured is laid off in squares, and field elevations taken at all intersections, as shown. The accuracy of the method depends upon the size of the squares. Contours are interpolated between known elevations in the same manner as previously described under Contour Sketching from Control Points.

This method is suitable to ground having a uniform slope, and if used where slopes are uneven and irregular, the size of the squares must be reduced until the lines between intersections have an even slope.

Since the map distance for interpolation is always the same, *i.e.*, equal to the side of a square, this method readily lends itself to the use of the slide rule. For example the elevation of intersection d-4 is 25.7, of d-5 is 26.2, the difference of elevation being 0.5; therefore the 26 contour is two-fifths of the map distance from d-5 to d-4, measured from d-5. For points having little difference of elevation, the interpolation for most surveys is sufficiently accurate if done by eye.

To Construct a Side Elevation from a Contour Map.

Figure 62 shows a side elevation constructed from the contour map directly above it. The result is nearly the same as a true perspective, and somewhat similar in outline to a photograph taken with the line AB as the picture plane.

To construct, proceed as follows:

1. Draw horizontal lines parallel to AB, spaced at a vertical distance apart equal to the contour interval. These lines should include the highest and lowest elevation shown on the map.
2. Draw projection lines perpendicular to AB and tangent to the respective contours.
3. These tangent points show the limits between visible and invisible points on the map, when viewed perpendicular to AB.
4. Extend projection lines to intersect the horizontal line having the same elevation as the contour to which the projection line is tangent.
5. Connect points thus projected and the result is an outline of the visible portion of the map.
6. Streams and gullies may be represented, if visible, by projecting their intersection with contours.

Figure 63 shows a portion of a contour map upon which a road location AB has been superimposed. The full lines represent the contours after the road has been graded, and the dotted lines show the portions of the original contours that will be changed by construction.

The road is to be 30 ft. wide and to be built at a -0.6 per cent grade starting at A at elevation

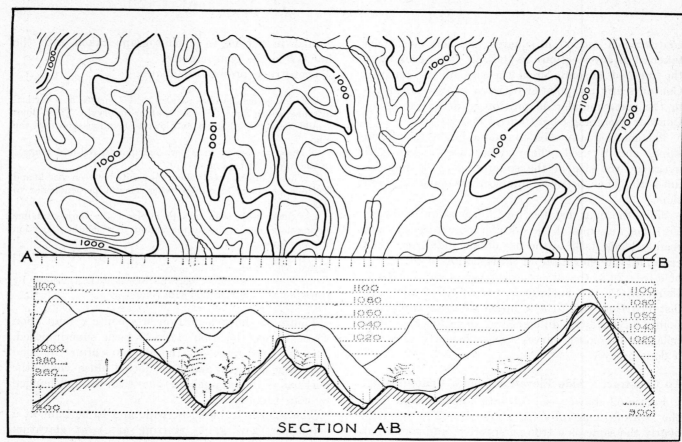

Fig. 62.—Side elevation from contour map.

1,054.70, side slopes one and one-half horizontal to one vertical. Scale 1 in. = 100 ft.

It is desired to construct a profile of the proposed road and draw in the contours showing the cuts and fills. The construction of a profile of a given line on a topographic map has been shown in Fig. 57 and need not be repeated here.

To draw in the contour changes, proceed as follows:

1. Figure where the new contours will cross the center line of road; 1,054 contour being 0.7 ft. below A would be $\dfrac{0.7 \times 100}{0.6} = 116\tfrac{2}{3}$ ft. from A. The remaining contours will be $333\tfrac{1}{3}$ ft. apart, as a 0.6 per cent grade falls 2 ft. in a distance of $333\tfrac{1}{3}$ ft. The road is assumed to be level transversely and these contours would cross the road at right angles to its center line.

2. Draw lines parallel to and 15 ft. on either side of AB. These lines show where the slopes of cuts and fills begin.

3. The directions of the contours on cuts and fills are found in the following manner: To locate the 1,060 contour on either side of Sta. 0 + 00, subtract the elevation of Sta. 0 + 00 from 1,060, and multiply the difference by 1.5. This gives (1,060 − 1,054.7) 1.5 = 7.95 ft. plus 15 ft. as the horizontal distance to be scaled from Sta. 0 + 00 at right angles to line AB to locate a point on the 1,060 contour. Similarly, the 1,054 contour is 6 ft. lower than 1,060, and, since the side slopes are 1.5 to 1, the 1,060 contour can be located opposite, where the 1,054 contour crosses the center line by scaling 15 (ft.) + 6 × 1.5 = 24 ft. at right angles to AB.

4. The two points thus located will give the location and direction of the 1,060 contour, which should be drawn to intersect the mapped location of the same contour.

5. Since a trimmed slope is a plane, all contours in the cut will be parallel and lie at a distance apart equal to 1.5 times the contour interval, which in Fig. 63 is equal to 3 ft.

6. If the contour crossings do not furnish enough interpolation points, others may be figured for any point on line AB.

Route Location.

Where good contour maps are available, many problems in highway and railroad location may be studied without recourse to a field survey. The conditions of maximum grade and curvature being

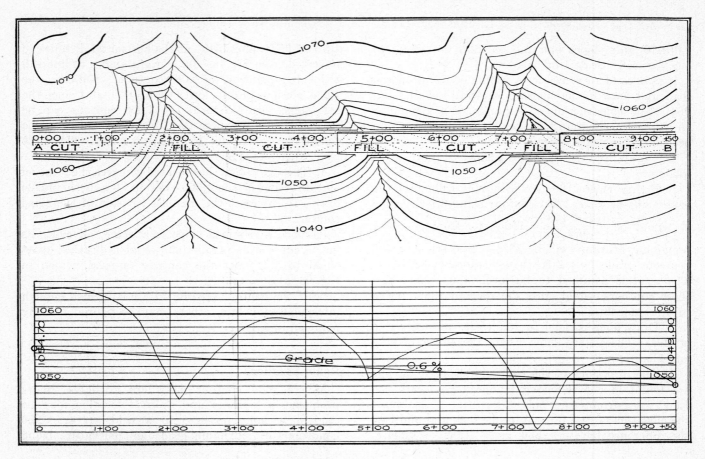

Fig. 63.—Plan and profile of highway.

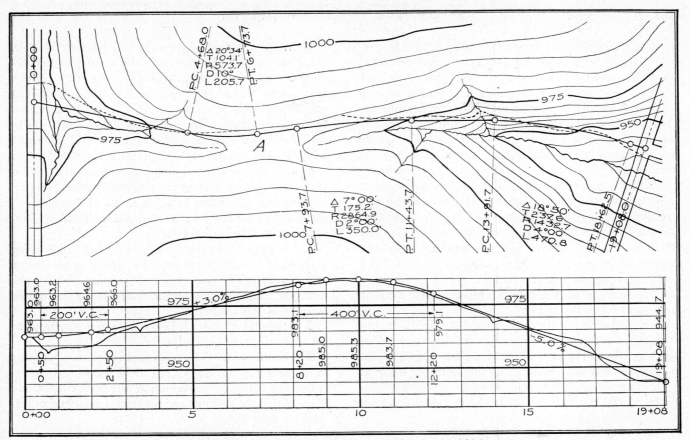

Fig. 64.—Plan and profile of paper location for a proposed highway.

known, a line of a given grade (called a "grade contour") may be located on the map, and the route located as nearly in accordance with the grade contour as the condition of maximum curvature will allow. A profile is then constructed by scaling the abscissas from Sta. 0 to a given contour and plotting the contour elevation as the ordinate. The grade line can then be adjusted to the profile, cross-sections constructed from the contours, and approximate calculations of earthwork made. Figure 64 shows a highway located across a divide and connecting two existing roads. In locating over a divide, it is best to start at the lowest contour on the divide (the 985 contour) and locate a grade contour down the ravine on either side of the saddle marked A. To locate a grade contour:

1. Select the starting point on the 985 contour near one end of the saddle.

2. Set a pair of dividers at a distance in inches equal to the contour interval divided by the scale of the map times the percent grade. The map scale in Fig. 56 is 200 ft. = 1 in., and the contour interval 5 ft. Then $5 \div 200 \times 0.03 = 5/6$ in. or a map distance of 166 ft., 8 in. This is the distance in which a -3 per cent grade will fall 5 ft.

3. With the dividers set at $5/6$ in., start at the point selected on the 985 contour and step the dividers to the next lowest contour toward Sta. 0. Connect the points where the dividers rest on the contour and repeat the operation until the contour nearest Sta. 0 is reached. The resulting line, shown dotted on Fig. 64, is a 3 per cent grade contour and represents a line on the ground surface having a 3 per cent grade.

4. A 5 per cent grade contour is similarly located from the 985 contour to the contour nearest the road intersection where the location ends. The spread of the dividers for the 5 per cent grade contour is $1/2$ in., or a map distance of 100 ft.

5. Tangents are now drawn, following the location marked by the grade contours as closely as conditions of topography and curvature will permit. The angles between tangents are scaled, and curves calculated to connect the tangents.

6. A profile of the projected location can now be drawn and grade lines adjusted to fit.

The results obtained by this method do not compare in accuracy with an actual field survey, but are extremely valuable in comparing alternate routes, and often eliminate preliminary surveys that would otherwise be necessary.

Visibility Problems.

To determine the points or areas invisible to the observer from any fixed point on a contour map, proceed as follows: In Fig. 65, taking the position of the observer as A and the line of visibility as AG, construct a profile of AG by projecting the contour crossings on AG to their respective elevation lines on the profile.

Interpolate summits and valleys and connect these projected points into a profile.

From the observer's position a on the profile, draw tangents to the summits at b, d, and f and extend these lines until they pierce the forward slopes at c, e, and g.

The portions of the profile invisible to the observer at a are from b to c, d to e, and f to g. In Fig. 65 there are three areas invisible from A, only one of which (the one subtending the line FG) is shown as a shaded area on the contour map.

To determine the invisible area draw lines radiating from A and tangent to all contours on the ridge line passing through F.

Extend these lines and determine their piercing points with the forward slope.

A line connecting all tangent points will be the boundary of the invisible area nearest the observer, while a line connecting all piercing points will define its boundary on the forward slope. The line ATP with its tangent point at T and its piercing point at P serves as an illustration.

Adjusting Road Surfaces by Use of Contours.

Figure 66 shows two highways intersecting on a superelevated curve. The control elevations for the paved surface are the center-line elevations taken from the profile plan and the edge elevations at the curbs taken from the superelevation tables. The primary problem is to construct a contoured surface passing through all these control elevations. This is done as follows:

1. Plot the intersection to a large scale, 1 in. = 10 ft. or larger.
2. Select a contour interval small enough to allow the scaling of elevations to at least 0.02 ft.
3. By cross interpolation locate the 1-ft. contours.
4. Space in all the intermediate contours by careful interpolation.
5. Elevations for any section at right angles to the center line may be scaled from the map at intervals as close as may be desired.

In Fig. 66 elevations on Section AA have been scaled to the nearest 0.01 ft. as an illustration. Any

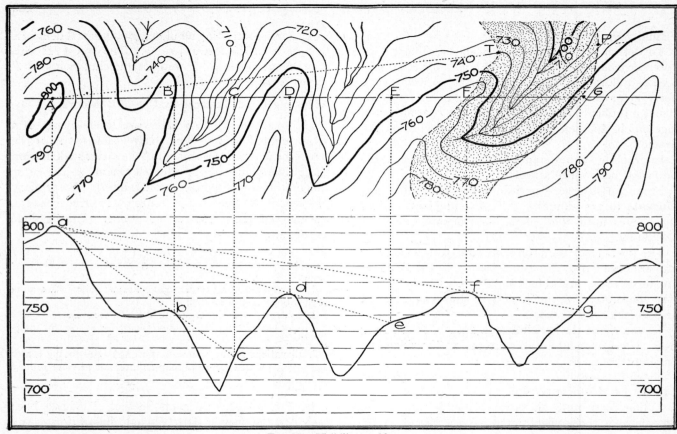

Fig. 65.—Visibility problems.

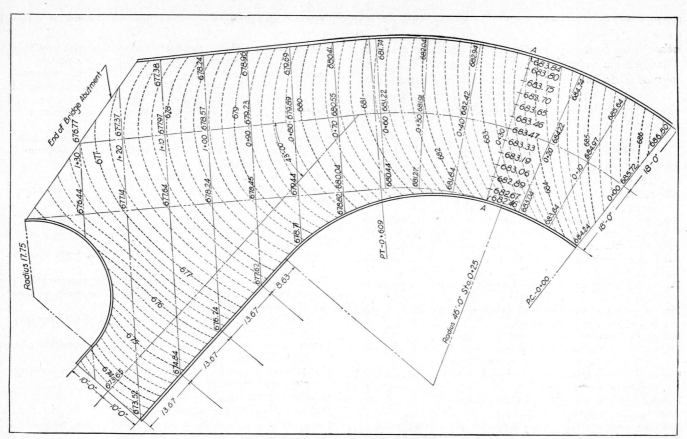

Fig. 66.—Contour adjustment of highway surfaces.

other section may be determined in the same manner. This graphic method may be used to adjust similar warped surfaces and avoid much tedious computation.

It is also valuable to the construction engineer in that he may determine by scale the correct elevation for any point on the intersection.

GLOSSARY OF TOPOGRAPHIC FORMS

Alkaline flat: The bed of a dried-up saline lake, the soil of which is heavily impregnated with alkaline salts.
Alluvial fan: The fan-shaped deposit of a stream where it issues from a gorge upon an open plain.
Amphitheater: A cove of glacial origin near the summit of a high mountain and nearly surrounded by the highest peaks.
Arroyo: A small, dry, gully or channel carved by water.
Atoll: A ring-shaped coral island nearly or quite encircling a lagoon.
Badlands: Waste or desert land deeply eroded into fantastic forms.
Bar: An elevated mass of sand or alluvium deposited on the bed of a stream, lake, or sea.
Barranca: A rock-walled impassable canyon.
Basin: An extensive depressed area into which the adjacent land drains and which has no surface outlet.
Bayou: A lake or intermittent stream formed in an abandoned channel of a river.
Bench: A strip of plain along a valley slope.
Bluff: A high, steep bank or low cliff.
Box canyon: A canyon having practically vertical rock walls.
Brow: The point at which a gentle slope changes to an abrupt one.
Butte: An isolated hill or small mountain with steep precipitous sides. The top may be flat, rounded, or pointed.
Canyon: A gorge or ravine of considerable dimensions; a channel cut by running water, the sides of which are a series of cliffs rising from the bed of the stream.
Cascade: A short, rocky declivity in the bed of a stream over which the water flows.
Cataract: A cascade in which the fall has been concentrated into one sheer drop; the volume of flow is usually much greater than that of a cascade.
Cave: A hollow space under the surface of the earth.
Cavern: A large, natural underground cave or series of caves.
Cay: A key, a small, low coastal island of sand or coral.
Cirque: A deep, steep-walled, amphitheaterlike recess in a mountain caused by glacial erosion.
Cliff: A high and very steep declivity.
Cone: A low conical hill built up from fragments ejected from a volcano.
Coulee: A cooled and hardened stream of lava; also sometimes applied to a wash or arroyo having an intermittent flow of water.
Cove: A small bay; also the abrupt heading of a valley in a mountain.
Crater: The cup-shaped depression marking the position of a volcanic vent.
Crevasse: A fissure in a glacier; also a break in a levee or stream embankment.
Cuesta: A tilted plain or mesa top.
Defile: A deep and narrow mountain pass.
Delta: An alluvial deposit at the mouth of a river, frequently in the shape of the Greek letter delta.
Desert: An arid region of any dimension.

Dike: A ridge having for its core a vertical wall of igneous rock.
Divide: The line of separation between drainage systems.
Dome: A smoothly rounded rock-capped mountain summit.
Drainage area: The watershed of a given stream or river system.
Drumlin: An elongated or oval hill of glacial drift.
Dune: A hill or ridge of sand formed by the winds.
Escarpment: A high steep face of rock of considerable length.
Estuary: A riverlike inlet or arm of the sea.
Everglade: A tract of swampy land covered mostly with tall grass.
Fault: A fracture in the earth's crust accompanied by a displacement of one side of the fracture in relation to the other.
Fan: A conical talus of detrital material.
Fork: One of the major branches of a stream.
Fumarole: A spring that emits steam or gaseous vapor.
Gap: An opening between hills or in a ridge or mountain chain.
Geyser: A hot spring the water of which is expelled at intervals by an accumulated volume of steam.
Glacier: A large stream of ice flowing down a valley or slope.
Gorge: A very rugged and deep ravine.
Gulch: A shallow canyon with smoothly inclined slopes.
Hanging valley: A high glacial valley that joins a more deeply eroded valley at an elevation some distance above its bed.
Hogback: A steep-sided ridge having a parallel trend to the adjoining mountains.
Hummock: An elevated piece of ground arising out of a swamp.
Kame: A small hill of gravel or sand made by a glacier.
Knob: A prominent peak with a rounded summit.
Lateral moraine: A low ridge of glacial drift formed along the side of a glacier.
Levee: An artificial bank confining a stream channel.

Marsh: A tract of low, wet ground.
Mesa: A flat-topped mountain bounded on at least one side by a steep cliff.
Moraine: Any accumulation of drift deposited by a glacier.
Muskeg: A bog or deep marsh.
Notch: A short defile or gap through a hill or mountain.
Oasis: A green or fertile spot in a desert.
Palisade: An extended rock cliff rising steeply from the margin of a stream or lake and of columnar structure.
Pass: A depression in a mountain range through which a road may pass.
Peak: A mountain with a single conspicuous summit.
Peneplain: A land surface that has been reduced to one of lower relief by water erosion.
Rapid: A short stretch of steeper slope between two relatively level parts of a stream bed.
Saddle: A shallow gap on a ridge.
Savanna: A grassy plain composed of moist, fertile land.
Sink: A depression in the land surface usually drained by a connection from its center to underground waters.
Slide: The exposed surface left in the trail of a landslide.
Spur: A sharp projecting ridge from the side of a hill or mountain.
Summit: The highest point of any undulating land.
Swale: A marshy depression in generally level land.
Talus: Fallen and disintegrated material forming a slope at the foot of a cliff or steeper declivity.
Terminal moraine: Irregular ridgelike deposits of alluvial drift left by a glacier as it melts and recedes.
Water gap: A gap through a mountain occupied by a stream.
Watershed: The ridge or summit separating two drainage systems; also, the area drained by a stream.
Wind gap: An elevated gap not occupied by a stream.

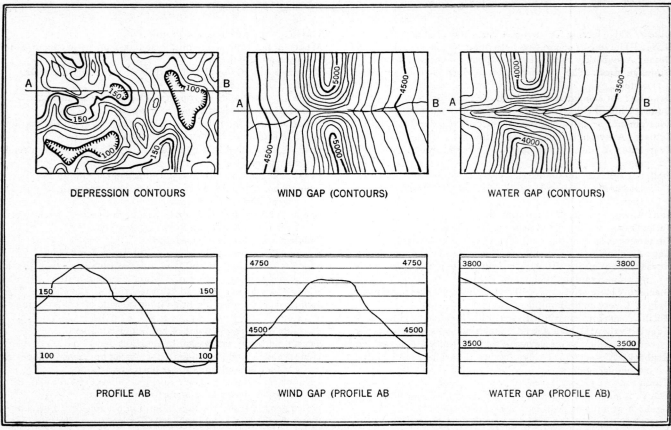

Fig. 67.

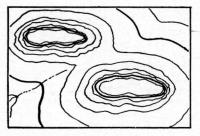
EROSION CYCLES (Contours)

LANDSLIDE (Contours)

SADDLE (Contours)

EROSION CYCLES

LANDSLIDE

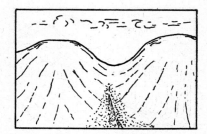
SADDLE BETWEEN TWO HILLS

Fig. 68.

HANGING VALLEY (CONTOURS)

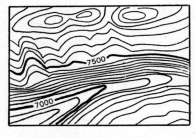
CIRQUE AND TROUGH (CONTOURS)

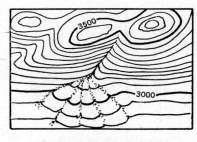
ALLUVIAL FAN (CONTOURS)

HANGING VALLEY

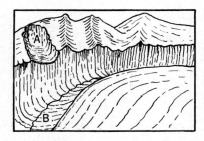
(A) CIRQUE (B) GLACIAL TROUGH

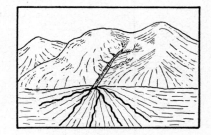
ALLUVIAL FAN

Fig. 69.

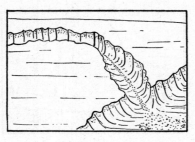
ARROYO

DESSERT SAND DUNES

FORMATION OF MORAINES

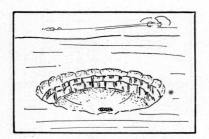
LIMESTONE SINK

MATURE VALLEY

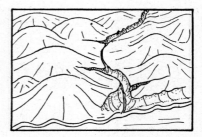
YOUTHFUL VALLEY

Fig. 70.

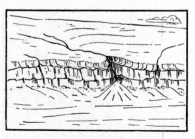  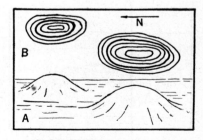

Fig. 71.

CHAPTER VI

COPYING—DUPLICATION—REPRODUCTION

The great majority of topographic drawings are either duplicated by printing on process paper to the same scale, or reproduced to a different scale by one of the photolithographic methods in common use. Since the draftsman creates the original from which the copy or duplication must be made, it is important that he have some knowledge of the various methods of reproduction.

Notes on Copying.

Simple line drawings are often transferred to opaque paper by laying the original drawing face up on the copy sheet and pricking the points through with a fine-pointed instrument. If many points are to be transferred, they should be numbered in pencil, both on the original and on the copy, as they are pricked through. This method is suitable only to line drawings, such as a plat showing a boundary survey.

Pencil Transfer.

To transfer any part of a topographic drawing to the same scale, proceed as follows:

1. Lay a piece of tracing paper over the original and trace the part to be transferred, in pencil.
2. Turn the tracing over and retrace all lines with a soft pencil (HB or B) sharpened to a fine conical point.
3. Turn tracing paper over, orient to its true position on the map, and tack down smoothly to the drawing board.
4. Transfer the penciled outlines to the map by rubbing over the pencil lines with some smooth, rounded instrument. A piece of tracing cloth with the smooth side up should be held between the rubbing instrument and the paper, to protect the paper.
5. Do not rub too hard, or you may move or stretch the tracing paper from its correct position.
6. The transfer should show a fine, clean line. If carefully done, it is very accurate.

Use of the Glass-top Table.

Another device for copying drawings on opaque paper is the glass-top drawing table or board. The drawing board consists of a piece of plate glass set flush with the top of the board and fitted with movable lights beneath the glass. These lights may be moved to any position under the drawing, thus preventing shadows. More elaborate fittings, such as a double

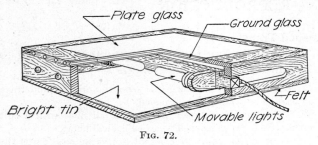

Fig. 72.

glass, with air space between to prevent overheating, and a bottom lined with bright tin to reflect all light upward, are used in offices where much copying is done. Figure 72 represents a type of glass board in common use. Delicate drawings may be copied on the heaviest paper by the use of this device.

Use of the Pantograph.

The pantograph may be used to copy any type of drawing either to the same scale, or reduced, or enlarged. Figure 73 shows a small inexpensive pantograph for drawing-board use. Figure 74 shows a pantograph most commonly used in drafting offices

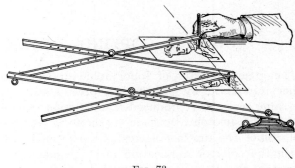

Fig. 73.

where the amount of work does not justify the expense of a suspended pantograph. Figure 75 shows a metal

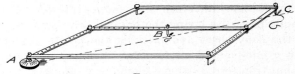

Fig. 74.

suspended pantograph used for accurate engineering work. The instrument is composed of four flat pieces of wood or metal joined together in such a way as to form a parallelogram. One of the points (A, B, and C, in Fig. 74) is fixed, and the other two movable. The movable points are so attached to the instrument that they will trace out exactly similar figures. There are several different forms of this instrument, but they are all based on the same principle. The two essential conditions are that the points A, B, and C of Fig. 74 must lie in a straight line, and that each point must be attached to one of three different sides (or sides produced) of the jointed parallelogram. Since any one of the three points may be fixed, and the other two movable, it is evident that by changing the relative positions of these points by moving them up or down the scale on the arms of the parallelogram, always keeping them in a straight line, the scale of the copy may be made to bear any desired relation to the scale of the original drawing.

These instruments are fitted with scales for setting the movable points, but because of imperfections in design, it is best to check the setting by trial on proportionate squares.

Proportional Squares.

A drawing may be copied either as a reduction or enlargement by the method of proportional squares.

Figure 76 illustrates the method of enlarging a topographic map.

1. Rule the drawing to be copied into squares of convenient size. Letter one set of coordinate lines and number the other.
2. Rule the copy sheet into squares whose sides are the same map distance as the original, but plotted to the scale desired in the copy.
3. Letter and number the coordinate lines to correspond to the original.

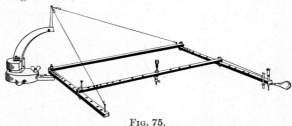

Fig. 75.

4. Set the pivot of a pair of proportional dividers, as shown in Fig. 77, so that the distance between points at opposite ends of the dividers is proportional to the map scales of original copy.
5. Transfer points from original copy with the dividers by their relation to the coordinate lines.

Tracing Paper or Cloth.

The most common method of copying drawings is by tracing the lines of the original in India ink on a transparent paper or cloth. Record or master draw-

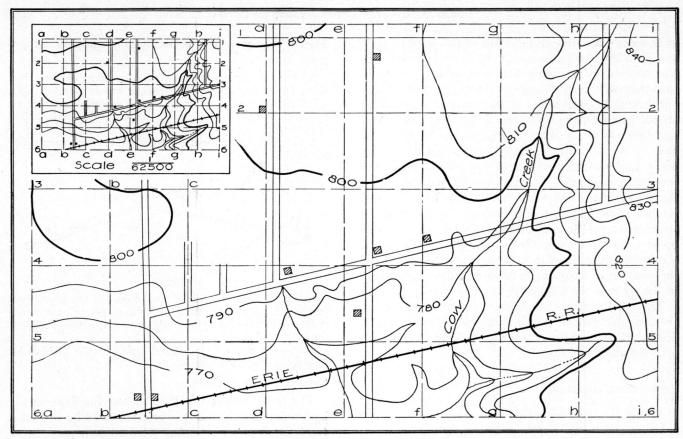

Fig. 76.—Enlargement by squares.

ings should be traced on cloth while drawings of a temporary nature may be traced on one of the thin transparent papers used for this purpose. Tracing papers lack the strength and stability of cloth and are inclined to become brittle with age. Tracing cloth is a fine-thread linen cloth, sized and made transparent with a starch preparation. The smooth side is intended by the makers as the right side, but most draftsmen prefer to work on the dull side, principally because it will take pencil marks well. Tracings are often inked on the smooth side to permit shading in pencil or color on the reverse or dull side. To trace a drawing, observe the following precautions:

1. See that the tracing cloth is stretched smoothly and evenly over the pencil drawing, as otherwise it may wrinkle and bulge.

2. Dust thes ide to be inked with chalk dust, talcum powder, or prepared pumice, and rub off with a clean cloth. This is necessary to remove a waxy film which prevents the flow of ink.

3. If blue prints are to be made from the tracing, the lines should be inked heavier than for a similar drawing made on white paper; because the contrast of a white line on a blue background is not so great as a black line on a white background.

4. If pencil work is done directly on the cloth, do not use a dull, soft pencil, as this prevents an even flow of ink.

5. Red ink will print much fainter than black, and should not be used, unless it is desirable to make certain lines inconspicuous. Blue will not print. Brown prints fairly well, but is not so sharp and clear as black.

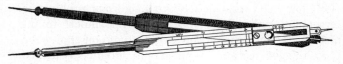

Fig. 77.

6. Tracing cloth is very sensitive to changes in humidity and temperature; a large drawing should therefore be traced as rapidly as possible. If it is not possible to trace a map in one day, you should trace completed sections to some natural division line such as a stream or road, and readjust the tracing before finishing.

7. Water will ruin tracing cloth. In warm weather draftsmen should either work with their sleeves down or use a paper mat to shield the parts of the drawing not in use.

8. To clean a tracing of dirt and pencil lines, rub lightly with a soft rag or piece of waste moistened in benzine or gasoline.

9. Erasures are more easily made on the smooth side, but the dull side will stand the most erasing. To erase ink lines from a tracing, place a triangle or some other hard smooth surface directly beneath the line to be erased, rub with a pencil eraser and moderate pressure until erased. If the surface of the cloth has been damaged by too vigorous rubbing or scratching with a knife, burnish with a smooth round piece of talc or brush a thick coating of collodion over the damaged section. This should produce a surface that will take ink.

10. To loosen the ink before erasing, use a thin-bladed knife kept very sharp to lift or scale the ink from the cloth. Never try to remove ink by scraping with the knife, as this damages the cloth.

Reduction by Perspective Projection.

To reduce the four-sided area 1-3-4-2 shown in Fig. 78 from a scale of 1/12,000 to a scale of 1/24,000 proceed as follows: Select the point O at any convenient location, preferably outside the area, and draw the rays O-1, O-2, O-3, and O-4 from the point O to the respective corners 1, 2, 3, and 4 of the known area. The dimensional ratio between the scale of the area and the scale of the reduction is equal to one-half. Divide the rays from O to each respective corner according to this ratio. Point 1 of the original area will fall at 1″ on the reduction exactly halfway

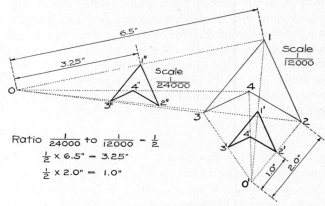

Fig. 78.—Reduction by perspective projection.

between O and 1. Draw line 1″-3″ parallel to line 1-3 and prolong to intersect with 0-3 at 3″. Transfer lines 3-4, 4-2, and 2-1 in a similar manner to complete the reduction.

Points 1″, 2″, 3″, 4″ should fall on the mid-points of rays O-1, O-2, O-3, and O-4 and may be checked by careful scaling. Reduction $1'$-$2'$-$4'$-$3'$ is equal in size and dimensions to figure $1''$-$2''$-$4''$-$3''$ but is constructed with the rays drawn from O' as a separate illustration.

This method of reduction is based on the theory of similar triangles, and its graphical accuracy largely depends on securing strong intersections between the lines of the reduction and the respective rays passing through each line. Strong intersections usually are found when the point O is placed nearly perpendicular to the longest line in the area to be reduced. This method may be used either to reduce or to enlarge a given figure.

Duplication.

The simplest method of duplication is a direct print from the original tracing on one of the many process papers now in use. The most common process paper used in drafting offices is blue-print paper. It is a white paper, free from sulphites, coated on one side with a solution of citrate of iron and ammonia, and ferricyanide of potassium. On exposure to light the portions of the tracing not covered with ink allow the light rays to produce a chemical action; which, when fixed by washing in water, gives a strong blue color. The portion of the print directly under the inked lines, not having been exposed to light, will wash out, leaving white paper. Blue-print paper or cloth may be bought ready sensitized in different degrees of weight and rapidity. When fresh, and of prime quality, it is a light yellowish-green in color and should wash out to a clear white before it has been exposed to light. Old or partially exposed paper is light grayish-blue in color, and will not wash white when immersed in water. The blue-printing industry has been so expanded and standardized that it is possible to get good commercial printing done at a reasonable cost, and few offices now do their own printing. To make a blue print, proceed as follows:

1. Lay the tracing in the frame with the ink side toward the glass.
2. Place the paper over it with its sensitized surface against the tracing.
3. See that you have perfect contact between paper and tracing, that no corners are turned down, and that both paper and tracing are free from wrinkles.
4. Expose to sun or artificial light a sufficient length of time to secure a good print. This time interval must be either determined by trial or fixed by your knowledge of the speed of the paper.

5. Upon removal from the printing frame, wash 5 to 10 min. in running water. An overexposed print may often be saved by prolonged washing.

6. The blue may be intensified and the white cleared by dipping in a solution of sodium bichromate containing $1\frac{1}{2}$ oz. of crystals per gallon of water. The prints must be well washed after this treatment.

7. Hang prints to dry where they are well protected from direct light rays.

Van Dyke Solar Paper.

The engineering profession has always felt the need for a positive black- or blue-line print on a white background. The first black-line prints were not entirely satisfactory, but the advent of the Van Dyke paper seems to have solved this difficulty. Van Dyke paper is a thin, sensitized paper which turns dark brown on being exposed to light, washed, and fixed in a bath of hyposulphite of soda. It is printed in the same way as a blue print, except that the tracing is placed in the frame with the ink lines next the sensitized surface of the paper. This results in a reverse print or "negative," and is not the final print. The inked lines appear as white, and the background a dark brown. The negative print is then placed in the frame, with the face of the negative next the sensitized side of the process paper, and the resulting print is a positive having colored lines on a white background. Any number of prints may be made from the Van Dyke negative, thus saving the wear and tear on the tracing.

Blue-print paper printed from the Van Dyke negative will give a blue-line print on a white background. Van Dyke paper will give a dark-brown line on a white background. Any of the process papers or cloths may be used with a Van Dyke negative to obtain a white print and colored lines.

The popularity of the above process is due to the ease with which penciled or inked changes may be made on the white background and its rapidly developing use for progress reports. The cost of black- or blue-line prints is about twice that of the ordinary blue print.

Sensitized Tracing Cloth.

To reproduce a tracing it is first necessary to make a Van Dyke negative print. Make a print from the negative in the usual way on sensitized tracing cloth. When sufficiently exposed, the tracing-cloth print is run through a water bath, then through a developing solution, after which it is hung up to dry. When dry, corrections can be made by erasing with a wet cloth. The entire surface is then gone over with a fixitive solution and allowed to dry. These tracing prints have the advantage of taking both pencil and ink lines, and are not affected by water or dampness as are the original drawings made on tracing cloth.

Photostat.

Duplication by photography has the added advantage over contact prints in that the reproduction may be either reduced or enlarged. The photostat machine is a large camera which photographs a negative image directly on a sensitized paper in place of using a celluloid film or glass plate. The sensitized paper is placed on a spool directly in front of the ground-glass plate and can be handled in much the same manner as the film-roll camera. To determine the size of reproduction within the limits of the machine, the operator measures the image on the ground-glass plate and compares with similar measurements on the original drawing. The negative is rephotographed to produce a positive print, and a reduction may be secured both on the negative and positive.

The Ozalid Dry-print Process.

This reproduction process uses light-sensitive papers or tracing cloth coated to produce blue, black, maroon, or sepia lines on a white background. These sensitized papers are subjected to direct positive-printing and are dry-developed by exposing the dry print to ammonia vapor.

This process has the advantage of speed and gives flexibility in the choice of papers and color of lines. Direct copies of record drawings may be made on transparent papers or tracing cloth and the reproduction, black-line on white background, used in printing additional copies. This saves much wear and tear on the record drawings.

For photographic reproduction for either enlargement or reduction it is first necessary to prepare a photographic negative, which in turn is projected onto a sensitized photographic film in order to produce a positive film known as a "diapositive."

This diapositive may now be used to secure contact prints on paper or cloth in any of the four colors desired.

Zinc Etchings.

When a great number of reproductions is required, and the expense of contact prints is prohibitive, maps are reproduced by one of the various methods of photolithography. The zinc-etching process consists, in brief, of making a photographic negative to the required scale, printing from the negative to a sensitized zinc plate, and etching and mounting the plate to produce a positive print. The old wet-plate process is used in making the photographic negative because it gives the maximum contrast in tones. Described somewhat in detail, the process is as follows:

1. A piece of clean glass is flowed on one side with an albumen solution, and allowed to dry.

2. A collodion solution with certain chemicals added to increase its affinity for silver nitrate is flowed over the dried albumen.

3. The plate is now immersed in silver-nitrate solution for several minutes, and becomes sensitive to light.

4. It is now transferred in a light-proof holder to the camera and is ready for use.

5. The plate is now exposed to light reflected from the drawing, which has been previously tacked onto a vertical board parallel to the plane of the photographic plate and directly in front of the lens. Now with the exposure of the plate under way, the dense portions of the image are being built up by the action of the strong light reflected from the white background of the map. Since the black lines of the map reflect no light, corresponding parts on the sensitized plate will not be affected, and will become transparent when washed.

Stripping the Negative.—After the negative is finished and dried, it is stripped, that is, the collodion film is treated with certain substances which enable it to be peeled off and transferred, after turning over, to another piece of glass. This is to reverse the image in order to make the final print the same as the original drawing.

Printing the Zinc Plates.—The film side of the stripped negative is now placed directly against the sensitized side of a zinc plate and exposed to strong light. Since the negative was reversed, the only part of the sensitized plate affected by light will be the black lines of the original copy. After the printing has been completed, the zinc plate is removed from contact with the negative and is given a thin coating of etching ink on the exposed side. The plate is now dropped into a basin of water, which immediately dissolves the sensitive coating beneath the ink which has not been affected by light. The ink also lifts from the plate in these parts, leaving the metal bare. The insoluble parts of the coating, with their ink covering, remain on the plate, which is then dried and dusted with a resinous powder which amalgamates with the ink to form an acid proof or "resist."

Etching the Plate.—A weakened bath of nitric acid is now given the plate in a rocking tub. When the unprotected surface of the metal has been eaten down slightly, the plate is dried and dusted over in four directions with a red resinous powder called dragon's blood. The plate is then heated slightly, to melt the powder. The brushing embanks the powder against the raised edges of the lines where the heat melts it to form a resist, thus preventing the acid from undercutting the lines. Each time a plate is exposed to acid, dusted with dragon's blood, and heated as above, it is called a "bite." Usually, three bites are sufficient to give ample height to the printing lines.

Mounting the Plate.—The larger open spaces on the metal plate are now "routed" or deeply cut with a high-powered drill. This is to provide against clogging of printer's ink in the press. The metal is now mounted type high on a wooden block, and the edges planed to bring the plate within the limits of the column rules. The plate is now ready for printing, and several thousand copies may be printed from one plate.

Reproduction of City Maps.

Any method of reproducing city or cadastral maps should have the following desirable features:

1. The scaling accuracy of the original should be preserved.
2. The original should be left in condition for future use.
3. Satisfactory copies of good legibility and appearance should be available at the lowest possible cost.
4. The press plates used in printing copies must be capable of preservation for use in printing future editions.

The method developed by the R. H. Randall Company of Toledo, Ohio, is recommended as best meeting the above requirements and is a combination of photographic, engraving, and lithographic processes. Briefly described, the procedure is as follows: Upon completion in the field, each original is supplied with all of the lettering indicated by its accompanying information tracing. This is done by carefully pasting on small paper tags which are printed in the proper style and size of type. The original is then photographed. The wet-plate process is used, and the camera equipment should be such that a negative of the entire sheet, including the margins, may be secured in one exposure, without visible variation or distortion. Here again, fidelity to scale should be emphasized. In order to secure scaling accuracy, the ground glass used in focussing should be marked by means of the same templet that is used in plotting the boundaries and projection lines upon the originals. By bringing the marks upon the original into coincidence with those upon the ground glass, the scale of the negative is assured. Two negatives are usually made of the original. These are coated with an asphalt solution which acts as both an opaque and an engraving medium.

The coated negatives are then engraved, the lines or lettering for each color being successively cut or washed through the opaque solution, and, after use in making a press plate, recoated and made ready for the engraving of another color. Press plates are made by the photolithographic process, in which metal sheets (zinc or aluminum) are sensitized by covering with albumen-bichromate solution, and are then exposed to light through the engraved negative, and developed out ready for the press. The offset press, in which the actual impression is transferred from the metal plate to a rubber-covered cylinder, and thence to the paper, is employed.

Engraving and Printing U. S. Geological Survey Maps.

The topographic field sheets are first thoroughly checked, lettered, and inked. A photolithographic print is then made of the finished drawing and given to the engraver as a guide. On each of three copperplates a projection is made. The purpose of the three plates is for the separation of the three features of culture, contours, and drainage, in order to permit of their printing in three separate colors on the finished map.

The next step is to bring a print of the drawing to the copperplate to act as a guide for the engraver. The method is as follows: A photographic negative on a wet plate, before drying, is flowed over with a solution of glue, bichromate of potassium, and several other ingredients, and allowed to dry exposed to light. The film of the negative retains and is built up by this solution, while the clear lines of the negative are not. This results in the lines themselves having a certain depth, while the background film surrounding them is slightly in relief from the glass.

After drying, the negative is rolled in all directions with a roll of wax which adheres in a slight film of wax to the background in relief, but leaves the deep, clear lines free of wax. The plate being now thoroughly coated with a light film of wax, a sheet of celluloid is laid down upon it and burnished. This causes a film of wax to be transferred to the celluloid, leaving the lines which have been depressed intact.

The sheet of celluloid, having on it a waxy film broken only by the lines of the design, is laid carefully over the copperplates on which the projection and register lines have been cut, and, by burnishing, the wax leaves the celluloid in sufficient thickness to form a waxy film on the copperplate, except in the lines of design. The plate thus coated with wax will resist the action of acid, and the wax also prevents the plate from being etched or discolored, except along the lines of design where no wax has adhered.

The design of the map now shows in bright copper lines on a background of waxy film. The plates are now placed in an airtight box, and a current of air, driven through sulphide of ammonium into this box, carries with it fumes which attack the exposed lines of design, making them black or brown, depending on the length of the exposure. Two plates having been treated in this manner, the culture and lettering are engraved on the first plate, and the contours on the second. The copperplate on which the culture and lettering have been engraved is rolled over with a roll of wax, resulting in a film of wax, except where the deep-cut lines of culture have been engraved. A sheet of celluloid is laid down and burnished, and a transfer of wax film made to the third, or drainage, plate, where the acid fumes are applied. When the wax is removed, only the culture lines show on the third plate.

The contour plate is treated in a similar manner, and the contours transferred to the drainage plate. We now have the topography and culture stained in their proper position on the drainage plate, but to get accurate registration (the accurate fitting of the drainage lines to the contours), the drainage is traced from a photograph of the original drawing and transferred to the drainage plate as follows: This tracing is made by scratching on celluloid, with a point, all the streams, rivers, and shore lines. The scratched lines on the celluloid are filled out with flour of sulphur and laid down in careful register, with the stained lines of culture and topography. The drainage now being lightly stained on the third plate, all inaccuracies between the contours and the slight bends of streams shown on the original drawing are corrected, and the drainage plate engraved.

Printing of Proofs.—A copperplate is entirely rolled over with ink, carefully wiped off, and the copper polished in order to remove any fine ink scum that may remain. By this means, the lines of the engraving have been filled with ink, and the surface of the plate remains entirely clean. A sheet of paper, which has been previously dampened, is laid upon the copperplate and the plate placed in the press. A rotating steel cylinder presses the paper to the surface of the copperplate, causing the dampened paper to adhere to the ink and remove it from the fine lines of the engraving. The proofs are examined, corrections noted, and the proofs returned to the engraver for correction. The plate is then cleaned of all guide marks and scratches, and made ready for lithographic printing.

Making of Transfer Sheets.—The plate is now printed, similarly to the method used for the copperplate proof, on transfer paper, with a greasy, acid-resisting ink. The transfer paper is covered with a coating of starch and gum, upon which coating the print adjusts itself.

The ink does not enter into the pores or fiber of the paper at all, but adheres only to the coating, the purpose of the coating being to afford an easy release of the ink from the transfer paper to the lithograph stone. In making the transfer, the printer, having printed the plate with the greasy ink on the coating of the transfer paper, lays the paper, printed side down, on the lithographic stone, pulling it through the transfer press. This first course through the press pastes the paper firmly to the stone. The printer moistens the back of the paper, which is now uppermost. The paper absorbs the moisture and releases itself from the film of starch, and is then removed from the stone. The ink, being of a greasy nature, adheres to the stone.

Printing.—The three copperplates, having been transferred to stone, can be placed in the press and printed with whatever color the rollers of the press are charged. The usual order of printing is as follows:

1. Culture plate printed in black.
2. Contour plate printed in brown.
3. Drainage plate printed in blue.

Usually, all three stones are printed from the same press, in order to secure accurate registration, although three presses may be used at once on rush work or large assignments. Colored maps should be confined to as few colors as possible, since the map must be run through the press once for each color used.

Mapping for Reproduction.

Maps drawn for reproduction by one of the photo-mechanical processes already described should be made on smooth white paper or tracing cloth. Black drawing ink should be used, and it is desirable that the original be larger than the required reproduction. Since it is always best to preserve the hand-drawn appearance of maps, the reduction should be slight, the best general size being one and one-half times linear, *i.e.*, the linear dimensions of the original are one and one-half times the length of the same line on the reproduction. If it is required to reduce a map to three or four times linear, care must be taken to make the original coarse in appearance. The shading should be open and light, and irregularities in water lining, hill shading, and crop symbols more pronounced. Failure to observe this rule will give the reproduction a smooth, mechanical effect and destroy its hand-drawn appearance.

If lines are drawn too close together, the space between them will entirely close up in the reproduction, and will mar the effect. This is especially true of water lining. A rough idea of a drawing's appearance after reduction may be had by the use of a reducing glass, *i.e.*, a concave lens mounted like a reading glass.

Select a line on the original which measured, say, 1 in. Then move the glass away from the drawing until its image, when viewed through the glass, measures ½ in. The glass, when held at this distance, will give the general appearance of a one-half reduction.

Lettering.—The size and distribution of lettering on a topographic map, always a major problem, becomes much more difficult when the map is to be reduced.

1. The size of letter must be selected to fit the reduction.
2. Use open and extended letters and do not crowd.
3. Lower-case letters e, s and m, and figures 3, 5, 6, 8 and 9 must be kept open and distinct to prevent choking in the reduction.
4. Use a pen that will give a black even line and avoid thin or gray spots in the ink lines.

Type Letters.—Many map draftsmen use printed type letters of suitable size and style to designate all numbers and names used on the map. The names of cities, railroads, rivers, contour numbers, etc., are printed on thin white paper and pasted on the original drawing in the proper places. This does not interfere in any way with photographing the drawing or etching the metal plate and adds much to the final appearance of the reproduction.

Line Drawings.—The same general rules apply both to maps and line drawings, the most common size for line drawings being two times linear. If a smooth finished effect is wanted on the reproduction, the original should be three to four times as large as the copy. In marking drawings for reproduction, it is best to show dimension lines, and mark the actual size of the finished copy.

General Rules for Inking.

1. Use all-black ink.
2. A line should have a solid black appearance.
3. A line should never appear weak or gray in color.
4. Lines may be thin if they are black.
5. Never water line in blue, as blue lines will not photograph.
6. Red or brown ink will reproduce in a weaker line than the same features in black.
7. Use a smooth white paper or tracing cloth.
8. Observe rules given under Lettering in paragraph on Mapping for Reproduction.

CHAPTER VII

MAP PROJECTIONS

The subject matter of map projection is presented in outline form and illustrated graphically for the benefit of the student who has neither the time nor the inclination for a more complete study. A complete discussion of the mathematics of the various forms of projection is beyond the scope of this text.[1]

With the steady increase in the use of state-wide plane coordinate systems in surveying and mapping, which in turn are based on some type of projection, it is becoming increasingly important that the map draftsman have at least a rudimentary knowledge of the different types of map projection in common use.

For the ordinary topographical survey or map covering a comparatively small area, no attention need be paid to the effect of the earth's curvature and rectangular coordinates may be used. When very large areas are mapped, due regard must be paid to the fact that the earth is an oblate spheroid with an equatorial axis of 12,756,412.8 meters and a polar axis of 12,713,167.6 meters (Clark 1866), which means that it is very close in shape to a geometrical sphere and is generally so considered for most practical mapping purposes. The dimensions above given are in terms of a meter whose length is 3.28083333 ft.[1]

The only strictly accurate map of the earth or any portion of it would be a model in the form of a globe. Such a map would be most inconvenient and cumbersome. Furthermore the earth's spherical surface cannot be developed without some type of distortion, although this distortion can be reduced to a very

[1] For an excellent treatment of map projection see Deetz and Adams, *U. S. Coast and Geodetic Survey Special Publication* 68.

[1] C. H. Birdseye, *U. S. Geological Survey Bulletin* 809.

negligible amount on small areas. Recourse is therefore had to projecting the terrestrial surface in some manner upon the plane surfaces that are called maps. The term "projecting" as used in this sense refers particularly to the method used in arranging the meridians and parallels on the map, although some forms of projections are geometrically correct.

Necessarily such methods of mapping the earth's surface involve errors of various kinds and magnitudes. In comparing the relative merits of several types of projections for a given purpose, the following criteria should be among those considered:

1. The accuracy of their scales along meridians and parallels.
2. The accuracy of their representation of areas.
3. The accuracy of their representation of forms of those portions of the earth's surface covered by the map.
4. The facility of their construction.

From a geographical standpoint, the ideal map should have the following properties:[1]

1. It should represent the countries with their true shape.
2. The countries represented should retain their relative size in the map.
3. The distance of every place from every other place should bear a constant ratio to the true distances upon the earth.

[1] Deetz and Adams, *U. S. Coast and Geodetic Survey Special Publication* 68.

4. Great circles upon the sphere, *i.e.*, the shortest distances joining various points, should be represented by straight lines that are the shortest distances joining the points on the map.
5. The geographic latitudes and longitudes of the places should easily be found from their positions on the map, and, conversely, positions should easily be plotted on the map when we have their latitudes and longitudes.

Map Projections Based on Use.

The various types of map projections cannot be rigidly classified into exclusive groups because of certain characteristics that may be common to two or even more groups, but they may be roughly classified, based on the use to which the maps made by the various methods of projections are best adapted.

1. It might be desirable that various areas on the map be comparable as areas regardless of correctness of shape. In this case what is known as an *equal-area* type of projection might be used with a fixed scale. A unit area such as a square inch anywhere on such a map would represent a certain fixed proportion of the earth's surface.

2. On the other hand, it may be necessary to produce a map in which correctness of shape is more important than anything else, even though this might mean a changing scale over the map. This type of projection is particularly desirable on maps of coast

lines, containing a number of the smaller geographical features such as bays, capes, inlets, etc. Such a type of projection would be known as *conformal projection*.

3. It may be that directions are the important thing on a map. Maps of this type are increasing in importance every day because of their use for long-distance air and sea navigation. In this case we want a map that gives correct directions from at least one point on the map (its center, say) and directions as correct as possible from all other points on the map. Since a common definition of direction used in surveying is the word "azimuth," the map projection in which correctness of direction is essential is called *azimuthal projection*.

4. A map is often required to serve a dual purpose, and a compromise method of projection may have to be used. In a case of this type it would be desirable to use a method that would reduce errors of shape and area to as small an amount as possible.

From an inspection of the above rough attempt to classify map projections based on intended use, it must be evident that maps commonly used are based on a compromise of methods in such a way as to produce results that are correct or reasonably correct for one purpose and do not give too much error when considered from other standpoints.

Common Types of Map Projections Based on Methods of Construction.

Assuming that by the term "map projection" is meant the method of arrangement on a map of the parallels of latitude and the meridians of longitude, it will be found that most types of map projection will fall in one of the following types:

1. Stereographic projection.
2. Orthographic projection.
3. Gnomonic projection.
4. Some form of cylindrical projection.
5. Some form of conical projection.

In illustrating these projections geometric figures have been used to demonstrate the basic idea back of a projection, but it should be kept in mind that not all types of map projection are simple geometric projections. For example, the projection of Mercator and Lambert's conformal projection use forms of cylindrical and conic projection, respectively, modified in order to gain conformality.

Stereographic Projection.—Stereographic projection assumes the eye to be on the surface of the earth, usually at the pole of a great circle whose plane is the plane of projection or the plane of the map, the eye being in the opposite hemisphere from the surface to be mapped. In the case of a map made on a plane tangent to the earth at the North Pole with the eye at the South Pole, the result would be as shown in Fig. 79A. If the map was made on the plane of the equator, it would be as shown in Fig. 79B. This type

drawing by assuming the point of projection at infinity, with the projecting lines parallel to each other and perpendicular to the plane of the paper. The method for determining the radii of parallels is shown in Fig. 80, where the plane of projection is taken as tangent at the pole.

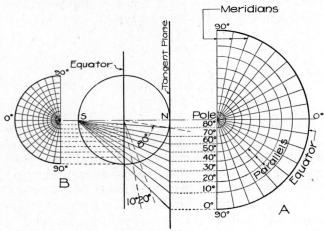

Fig. 79.—Stereographic projection.

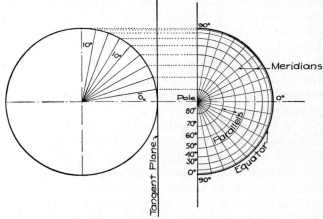

Fig. 80.—Orthographic projection (polar).

of projection produces a conformal but not an equal area map.

Orthographic Projection.—In orthographic projection the map is produced as in ordinary engineering

Gnomonic Projection.—This type of projection differs from stereographic projection in that in gnomonic projection the projecting lines radiate from the center of the sphere to the map, which is on a plane tangent to the sphere. As a result of this procedure all great circles are shown as straight lines on the map.

In Fig. 81 the tangent plane is the plane tangent to the earth at the pole, and it should be noted that a

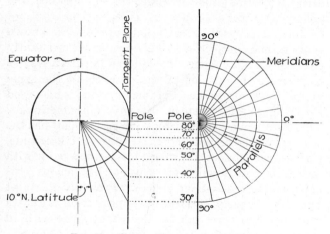

FIG. 81.—Gnomonic projection (polar).

complete hemisphere cannot be shown on any one plane, as in the case of the orthographic and stereographic projections, because of the increasing "spread" of the radiating projecting lines, which become parallel

to the plane of projection at 90 deg. Had the plane been tangent at the equator the meridians would have been straight parallel lines unequally spaced. The parallels would have been nonconcentric curves and

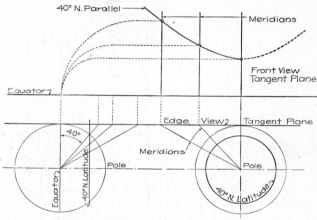

FIG. 82.—Gnomonic projection (equatorial).

the equator a straight line, as shown in Fig. 82, where the construction of one parallel is shown.

Cylindrical Projection.—The use of the developed cylinder as a device for mapping the earth's surface has long been in use. In this type of projection the map may be produced by projection on a right cylinder tangent to the earth at the equator, with the cylindrical elements parallel to the earth's polar axis. The cylinder is then developed or unrolled to produce

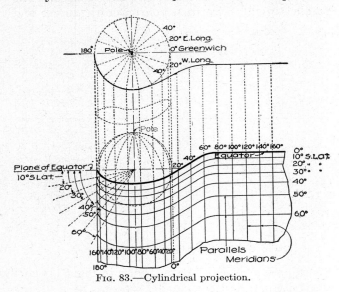

Fig. 83.—Cylindrical projection.

the map as shown in Fig. 83. When the meridians and parallels are projected onto the cylinder tangent to the earth at the equator, they appear as straight lines. In simple cylindrical projection the meridians of longitude extend as parallel, equally spaced straight lines from the equator to the pole, and the parallels of latitude as unequally spaced straight lines more widely spaced at the poles than at the equator. Disregard of convergence of meridians in simple cylindrical projection produces large errors in area near the pole. Mercator's projection is a special type of cylindrical projection and is one of the best known types of map projection. This projection constantly changes the scale along the meridian in such a way as to increase the spacing of the parallels at the same rate as the error produced in the longitude is increased by using parallel instead of converging meridians. This results in a map which at any given point has the same scale as regards both latitude and longitude, giving true shapes in small local areas and with meridians and parallels crossing at right angles as is actually the case on the sphere. Probably the most valuable property of Mercator's projection is that a course or line making a fixed constant angle with all meridians appears as a straight line on the map. This is known as a "rhumb line." Such a characteristic is exceedingly valuable to navigators in laying off their courses. Mercator's projection, while considered a cylindrical projection, is conformal and cannot be projected geometrically.

Figure 84 shows approximately the arrangement of the meridians and parallels in Mercator's projection. A variation of this type of map projection is known as "transverse Mercator" and may be looked upon as the ordinary Mercator's projection turned through 90 deg. so that the cylinder of projection is tangent to the earth along a meridian passing through the pole.

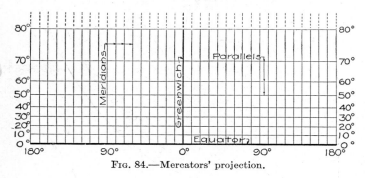

Fig. 84.—Mercators' projection.

This projection or some variation of it is used in the United States in connection with state-wide coordinate systems. As so used the transverse Mercator projection is based on a cylinder cutting the earth's surface in two circular arcs parallel to the central meridian of the map. The cylinder cutting these arcs is concentric with but slightly smaller than a tangent cylinder passing through the central meridian, as illustrated in Fig. 85. This cylindrical projection is modified (altered slightly) to give conformality and is not therefore a true geometrical projection.

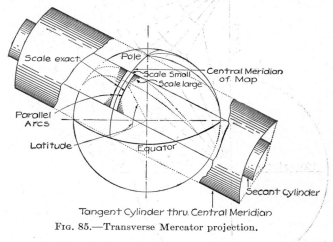

Fig. 85.—Transverse Mercator projection.

Conic Projection.—In this type of projection a cone tangent to the terrestrial sphere at the central parallel of the map may be used, with the apex of the cone falling on the earth's axis prolonged. The cone is then developed or unrolled to produce the map, as shown in Fig. 86. Instead of a central projection

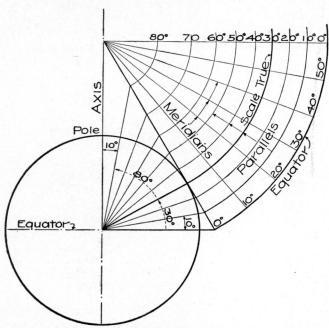

Fig. 86.—Conic projection (tangent cone) central perspective projection.

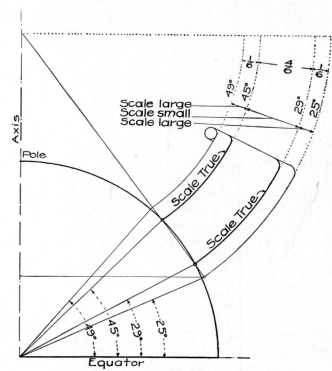

Fig. 87.—Lambert conformal projection (secant cone).

being used for determining the spacing of the parallels along the central meridian, as in Fig. 86, the scale of the spacing of the parallels and meridians may be made the same as that actually on the sphere. This results in the parallels being represented on the map by a series of concentric circles at their true distance apart with the central meridian straight and the other meridians slightly curved towards it and spaced as on the sphere. Bonne's projection is an example of this type.

Another modified conic projection is the well-known Lambert conformal, in which a cone is passed through two standard parallels and is then developed. While these two standard parallels are true to scale, they do not coincide exactly with those of the earth, because they are spaced to provide correctness of shape so that a map based on such a projection has a changing scale. However, the Lambert conformal projection produces comparatively slight distortion, particularly in east and west directions. Another variation of conic projection is polyconic projection, in which a series of cones is used, one cone for each parallel of latitude, all being developed from a central meridian. This type of projection is largely used in the United States, particularly by United States Government agencies such as the U. S. Coast and Geodetic Survey. Figure 87 illustrates the intersecting-cone or Lambert conformal projection, while

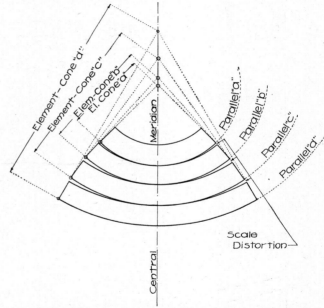

Fig. 88.—Developed polyconic projection showing scale distortion.

Fig. 88 shows the scale distortion in polyconic projection.

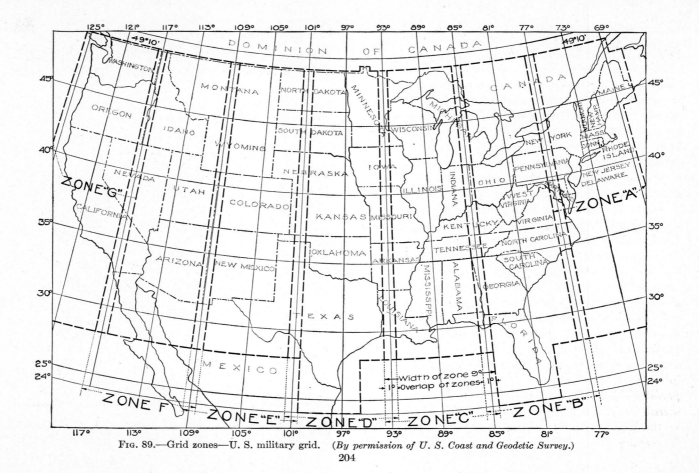

Fig. 89.—Grid zones—U. S. military grid. (*By permission of U. S. Coast and Geodetic Survey.*)

Military Grid System.

For military purposes, the entire United States has been covered with a grid system consisting of squares determined by rectangular plane coordinates. This system is superimposed on a basic polyconic projection divided into seven zones, each zone covering a range of 9 deg. of longitude with an overlap of 1 deg. of longitude with zones east and west. Each zone is given a letter, zone A being the most easterly and zone G the most westerly. The grid system is based on units of 1,000 yd. The grid system is matched with the polyconic projection of each zone by making the north and south grid line agree with the central meridian of the zone, the origin of each zone being the intersection of its central meridian with the 40°30′ N parallel of latitude. The coordinates of this point are assumed as $x = 1,000,000$ yd., $y = 2,000,000$ yd.

Owing to convergence of the meridians, grid north and true north agree only on the central meridian of a zone, and the scale error due to the use of polyconic projection amounts to about 0.2 per cent.

Figure 89 shows the entire grid system and Fig. 90 one zone, C.[1]

---

[1] Reprinted by permission of the U. S. Coast and Geodetic Survey.

Deetz and Adams in their excellent work on map projection published as *Special Publication* 68, by the

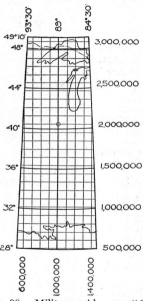

Fig. 90.—Military grid—zone "C."

U. S. Coast and Geodetic Survey, give the following comparison of errors of scale and errors of area in a map of the United States on four different projections:

Type of projection	Maximum scale error, per cent	Maximum error of area, per cent	Maximum error of azimuth angle
Polyconic................	7	7	1°56′
Lambert conformal conic..	2½*	5	0°00′
Lambert zenithal equal-area..................	1⅞	0	1°04′
Albers................	1¼†	0	0°43′

* With standard parallels at 33° and 45°; between latitudes 30½° and 47½° only 0.5 per cent. Strictly speaking these percentages are not scale error but change of scale.

† With standard parallels at 29°30′ and 45°30′.

Among many types of map projections, the following might be named:

1. Albers equal-area.
2. Bonne's.
3. Simple conic.
4. Simple cylindrical.
5. Gnomonic.
6. Lambert conformal.
7. Lambert zenithal equal-area.
8. Mercator's.
9. Transverse Mercator.
10. Orthographic.
11. Polyconic.
12. Rectangular.
13. Stereographic.

Lambert System	Transverse Mercator System
Arkansas	Alabama
California	Arizona
Colorado	Delaware
Connecticut	Georgia
Iowa	Idaho
Kansas	Illinois
Kentucky	Indiana
Louisiana	Maine
Maryland	Michigan
Massachusetts	Mississippi
Minnesota	Missouri
Montana	Nevada
Nebraska	New Hampshire
North Carolina	New Jersey
North Dakota	New Mexico
Ohio	Rhode Island
Oklahoma	Vermont
Oregon	Wyoming
Pennsylvania	Florida*
South Carolina	New York*
South Dakota	
Tennessee	
Texas	
Utah	
Virginia	
Washington	
West Virginia	
Wisconsin	
Florida*	
New York*	

* Both systems used.

Of these the most commonly used in the United States are Lambert conformal, polyconic, and transverse Mercator, the polyconic for country-wide use, the Lambert conformal for states with great east and west length, and the transverse Mercator for states with great north and south length.

The table on page 206 indicates the types of projection used as a basis for the plane coordinate systems of the various states in the United States.

CHAPTER VIII

SUGGESTIONS FOR OFFICE PRACTICE

The limited time devoted to topographic drawing in the curricula of most engineering schools leaves many items of practical information, which are of great value to the draftsman, to be gained through experience. This chapter makes no attempt at completeness, and includes only those commercial usages in topographic drawing which have come within the author's experience. It is hoped to stimulate in the student and draftsman an interest in accumulating and preserving as permanent notes all the practical information with which he may come in contact.

Instruments Not in Common Use.

The equipment used in making student drawings is generally as meager as actual necessities permit. The student, therefore, is not familiar with the use of many instruments common to commercial drafting rooms.

Straightedge.—Engineering maps are made with greater accuracy than much of the drafting work of allied professions. Furthermore, survey lines are not parallel or perpendicular to each other, except by chance, and must be plotted accurately in any

FIG. 91.

direction the notes require. All maps are therefore laid out starting from a long, straight line drawn on the paper by means of a straightedge. Steel straightedges are more commonly used than those of wood, since they are more accurate and will not warp or become rough along the edges. They are made in sizes 3 to 6 ft. in length and beveled along one edge. In penciling, the beveled corner is up, thus permitting the sharp pencil to draw a line very close to the straightedge. In inking, the beveled corner should

Fig. 92.

be down, to prevent the ink's flowing under the ruling edge. Figure 91 shows a steel straightedge, while

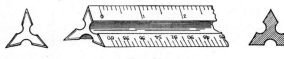

Fig. 93.

Fig. 92 shows a wooden straightedge with bakelite edges.

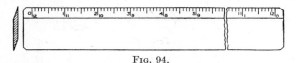

Fig. 94.

Scales.—It is assumed that the student is familiar with the engineer's and architect's scales, both in the flat and the triangular form (see Figs. 93 and 94). The student, however, is generally not familiar with the best practice in precision scaling. A few simple rules may therefore aid the beginner:

1. Keep the eye directly over the scale reading it is desired to plot.
2. Use a very sharp pencil or a needle-pricking instrument held vertical to mark the point.
3. Check the scale reading at both ends of the scaled distance and be sure the scale does not move before the point is marked.
4. Check scaled distances by using some initial reading other than zero.
5. Scales graduated finer than 80 divisions per inch are very hard to read without the aid of a glass, and it is better to use some fractional scale of larger divisions and estimate the fractional part by eye. For example, to plot 100 ft. = 1 in., use a 10, 20, or 50 scale, depending on accuracy required. To plot 200 ft. = 1 in., use 20 scale; for 300, use 30 scale; etc.
6. For maps of large areas plotted to a small scale, specially constructed scales with decimal subdivisions should be used.

Parallel Ruler.—This is a steel straightedge beveled on both sides and mounted about $\frac{1}{32}$ in. high on two rollers of exactly the same diameter. This instrument is used for drawing parallel lines, and, when handled with care to prevent the rollers from slipping, can be made to do very accurate work. It is especially useful in drawing sidewalks, buildings, and other rectangular objects parallel to streets or roads.

It can also be used for shifting traverse lines to new positions, blocking out titles, stress diagrams in structural design, section lining, etc. A cut of this instrument is shown in Fig. 95. A folding ruler is shown in Fig. 96.

Fig. 95.

Proportional Dividers.—This instrument is similar to an ordinary pair of dividers, with the exception that the pivot about which the legs move is adjustable, and the legs extend on both sides of the pivot point.

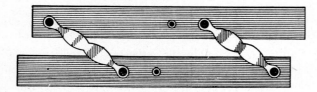

Fig. 96.

The pivot can be moved up or down the slot and the legs clamped in any desired position, thereby changing the relative length of the legs on opposite sides of the pivot. Proportional settings are marked on the scaled edges of the slot, but should be checked before using. The use of the instrument is confined to reducing or enlarging drawings without the use of a scale. For illustration, see Fig. 77.

Special Pens.—The *contour pen* shown in Fig. 97 is constructed similar to a right-line pen except that

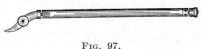

Fig. 97.

the blades of the pen are attached to a metal shaft inside the hollow handle, which may either be locked in position or allowed to turn on a swivel. The blades are shaped in a flat S curve, which allows them to drag behind and follow the direction in which the hand is moved. The free movement of the revolving shaft keeps them in correct position and prevents

Fig. 98.

side movement. Some practice is required before a draftsman can use this pen correctly, but when once skilled in its use, he will find it indispensable in drawing contours, shore lines, etc. A good contour pen can be adjusted to any weight line and give a uniformity impossible to secure with the ordinary pen. Figure 98 shows a *railroad pen* having two sets of blades mounted on a single handle in such a manner that the distance between the two sets of blades can be regulated by a tangent screw. It is useful in drawing roads, canals, railroads, curbing, and other details requiring two parallel lines with a small space between

Fig. 99.

them. They must be held almost vertically and used with very little pressure against the T square or triangle. They may also be used in drawing heavy ink lines by setting the blades to the required width of line and filling in the center portion. Figure 99 shows a *Swede* or *detail pen* used on large drawings. Its chief advantages are that it:

1. Carries more ink and will not clog easily.
2. Does not require filling often.
3. Is very suitable for borders and heavy lines.

Figure 100 is a *rivet pen* which consists of a bow pen mounted on a round handle fitted with a sharp point at its lower end. The pen has two motions with respect to the handle *viz.:* it slides up or down the handle, within limits, and it may be rotated around the handle. It must be held vertical when in use, the

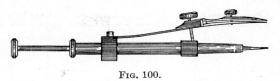

Fig. 100.

size of the circle being regulated by a tangent screw. Its chief use is in drawing small circles accurately

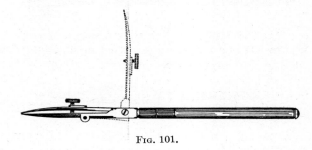

Fig. 101.

and speedily, such as rivet heads, traverse points, and symbols requiring small circles.

Figure 101 is a *jacknife pen*. One blade of this pen is hinged and made movable for cleaning purposes.

Figure 102 is a *border pen* having a third blade placed between the outer blades to prevent the too-rapid flow of ink when drawing a wide line.

Fig. 102.

Figure 103 is a *double-swivel pen* similar in operation to the contour pen. It is used in drawing parallel, freehand lines.

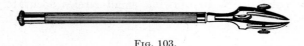

Fig. 103.

Figure 104 is a *crow-quill pen* used in sketching fine freehand lines on topographic maps.

Fig. 104.

Figures 105 and 106 are *Payzant pens* used for heavy or solid-stroke lettering. They may be purchased in a variety of sizes, and are useful in rapid lettering.

Beam Compass.—This instrument, shown in Fig. 107, is used for laying out circle arcs having a larger radius than can be set off with the ordinary instruments. It consists of a smooth strip of wood, similar in size and length to the ordinary yardstick, fitted

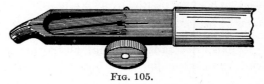

Fig. 105.

with two metal attachments which may be clamped in any position along the wooden strip. One of these

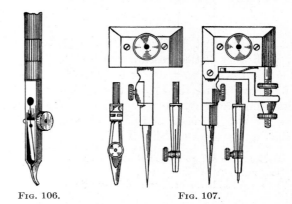

Fig. 106.   Fig. 107.

attachments is fitted with a needle point and the other with either a pencil or pen, with its attendant slow-motion screw for close settings. Some sets are also fitted with a spring rider, which consists of a spring mounted on two small wheels and arranged to clamp to the wooden strip near the point carrying the pencil or pen. This arrangement carries the pen free of the paper, except when a slight pressure is used. It also lifts the pen vertically from the paper, when the pressure is released.

Care of Instruments.

To remain in good condition and give long service, drawing instruments must be kept clean and free of dried and incrusted ink, and placed in a dry case when not in use. Pens and needle points must be frequently sharpened.

To clean a pen of incrusted ink, dip in ammonia water to dissolve the ink, wipe, and polish with a chamois skin or soft cloth. Instruments to be stored any length of time or used out of doors in a surveying camp should be coated with a light film of vaseline. Pens in constant use require sharpening, and every draftsman should be able to keep his own pens in good condition. A pen properly sharpened should be sharply oval in shape with the point of the oval symmetrical with the axis of the pen. It should never be ground to a V-shape or flat, broad oval, as neither of these shapes will ink properly. Many draftsmen wear their pens on one side only, and in sharpening such a pen the tendency is to get the point of the oval off center. The best stone to use is a hard, fine-grained Arkansas sandstone, which takes oil readily. Do not use a carborundum or a carpenter's oilstone, as they are too coarse grained for sharpening pens. To sharpen a pen, proceed as follows:

1. Screw the nibs together until they touch. Hold the pen in the same position as when ruling a line, drawing it across the stone by starting the stroke with the pen, making an angle of 25 to 30 deg. with the stone, and raising the top end of the handle gradually as the pen moves across the stone, until it has passed the vertical position. Repeat in both directions until the nibs have the proper sharp-oval shape, and are exactly the same length.

2. Open the nibs slightly and observe the ends of the nibs with a magnifying glass. They will be found to be very dull and have a flat bright spot at the pointed end.

3. Hold the pen at a slight angle with the stone and rub it back and forward across the stone with a rocking motion, to fit the shape of the blades. Repeat

until the bright spots on both nibs disappear, when viewed with a glass.

4. Test by drawing a line, without ink, across the paper. The blades should not be sharp enough to cut the paper. If they are, repeat the dulling operation, and resharpen.

5. Make the final test by drawing ink lines. They should be clean and sharp at any width down to the finest hair line.

6. Any roughness developing on the inner side may be removed by opening the blades wide and bringing the full inner surface in contact with the stone. In no case try to sharpen the inner side of the blades.

7. A few drops of 3-in-1 oil will prevent the stone becoming hard and glazed.

8. Do not experiment on a new pen. Try your hand on an old one first.

Use of Curves.

Landscape and city topography introduce a variety of irregular and reverse curves which produce an added burden to the draftsman who must ink the map. There are many mechanical aids in the way of irregular curves, with which the draftsman should become familiar.

Irregular Curves.—Figure 108 shows a partial set of curves used in a shipyard drafting office. Almost any irregular curve could be matched by curves chosen from this set. The curves are made of celluloid or metal and may be bought at any drafting-

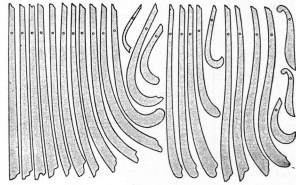

FIG. 108.

supply house, either singly or in sets. Figures 109 and 110 show a variety of shapes called "French curves." They are made of hard rubber, wood, or celluloid, and are used to guide the pencil or pen in drawing irregular curved lines on the map. Figure 111 shows a flexible curve which consists of a strip of rubber fastened to a flexible metal back. This curve may be bent or twisted to fit any irregular curve, and may be used as a guide against which the pen may be held in tracing it out.

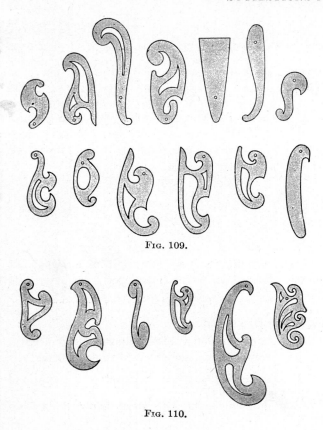

Fig. 109.

Fig. 110.

Fig. 111.

Railroad Curves.—For drawing arcs of curves of long radii, such as railways, highways, and city streets, curves made of hard rubber, wood, celluloid, or metal are used. These curves usually come in sets, marked as to scale and degree of curve. The most common scale is 1 in. = 100 ft., with sizes at 15-min. intervals from 0 deg. 15 min. to 15 deg. 0 min. Both edges of the curve are to the same radius, with a short straightedge tangent to the curve at one end. The point where the curve begins is marked by a line perpendicular to the tangent. An assembled set of railroad curves is shown in Fig. 112.

Tinting a Drawing.

The appearance and legibility of ink drawings may often be improved by tinting. This operation is preferably done before inking, but may be done afterward if strictly waterproof inks are used. The colors used may be either moist water colors or dilute colored inks. The paper must be stretched to prevent wrinkling, unless heavy papers, such as Bristol, are used.

The operation is as follows:

1. Dampen the paper with a sponge, until limp. Lay on a clean board or table face down, and dry about 1 in. of the outer edge with a blotter Brush glue or library paste on the dried edges, turn the paper over, and place on the drawing board. Rub the edges down until set, and place in horizontal position to dry.

2. Clean drawing of all dirt and pencil marks and proceed to apply the water colors as described in the chapter on Topographic Drawing in Colors.

If desirable, tracings may be tinted by inking on the smooth side and applying the color on the dull side. Tracings are best tinted by using colored crayons, as excess moisture will ruin the tracing cloth. To tint a tracing with water color, use very little color on a sable or camel's-hair brush, and work

Fig. 112.

rapidly. Small areas may be tinted in this way, but do not attempt to water color large areas on tracing cloth.

Mounting Display Maps.

Maps made on thin tracing paper are mounted in much the same manner as described under tinting.

1. Lay the drawing face down and brush a narrow strip of paste entirely around the edge.

2. Turn drawing face up, and moisten lightly with a damp sponge.

3. Smooth the drawing from the middle towards the sides, and bring the glued edges into contact with the mounting mat.

4. Stretch the paper gently by light pressure with the thumbs, on opposite edges.

5. Smooth out all wrinkles, set the glued edges, and place ... al to dry.

...cing cloth should be mounted with dry mounting tissue, such as that used by photographers in mounting photographs. It may be purchased in 5-yd. rolls 20 in. wide.

1. Cut a sheet of mounting tissue the exact size of the drawing.

2. Lay the tracing face down, and place the sheet of mounting tissue on the reverse side.

3. Fix in place by touching three or four spots with an iron just hot enough not to scorch.

4. Place in position on mat, and iron from the center toward the edges. Protect the surface of the tracing cloth with thin muslin, while smoothing into place with the hot iron.

Cloth-mounted Maps.

Drawings subject to much handling and rough usage are frequently mounted on cloth. The adhesives used are library paste and liquid glue, either of which gives satisfactory results, except that sheets mounted with glue are not so flexible as those mounted with paste. There are two methods in use, hot mounting, and cold mounting. The method used will depend somewhat upon the weight and quality of the paper, and the final purpose for which the maps may be used. The method of hot mounting is as follows:

1. A piece of first-grade white sheeting is stretched tightly and tacked down over a cloth-covered table.

2. Add a small amount of water to the paste and heat until it becomes clear. Apply paste to the back of the print with a broad, flat brush, working from the center toward the edges.

3. Allow $\frac{1}{2}$ to 1 min. for uniform expansion of the paper, and then place face up on the stretched cloth. Iron print from center toward the edges with a circular motion, until edges are stuck. During this operation the iron should be just hot enough not to scorch, and be well paraffined,to work smoothly over the print.

4. After once over with the iron, remove tacks and iron until dry. Never iron the cloth side, as this will cause steam blisters under the print.

Mounting with cold paste is very similar:

1. The cloth is stretched and print primed with paste, as described under hot mounting.

2. A heavy print roller is used in place of the hot iron, to attach the print to the cloth.

3. The mounted print must be left to dry in its mounted position before the tacks may be removed.

Hot Mounting Thin Paper.—Thin paper is very hard to handle without tearing or wrinkling after the paste has been applied, hence some mechanical means must be used to keep the print even and straight during the ironing process. The writer has seen several methods used, but believes the following device best suited to an amateur:

1. Place several thicknesses of newspaper under the stretched cloth, since the hot paste must be applied to the cloth, instead of to the thin paper.
2. Roll the print, face in, on a roll of detail paper and place at one end of the stretched cloth.
3. Apply hot paste, beginning at this end, unroll the print as the paste is applied, and follow up as fast as the print is unrolled with the hot iron.
4. Finish by ironing gently until dry.

Cold Mounting Thin Paper.—Thin prints may also be cold mounted on "Hollister photo cloth," which may be purchased in 10 yd. rolls, 36 in. wide.

1. Roll the print, face in, on a roll of detail paper, as in hot mounting.
2. Run the Hollister photo cloth through a tray of water, and place, sticky side up, on a large glass plate.
3. Unroll print slowly, starting at one side and squeeze to cloth as fast as unrolled.
4. Finish by rolling with a heavy photoprint roller.

Mounting with Rubber Cement.—Prints subjected to moisture conditions of outdoor work should be mounted in rubber cement as follows:

1. Both the back of the print and the stretched cloth or mat upon which it is to be mounted are given a thin coating of rubber cement, applied rapidly with a brush.
2. The cement is allowed partially to set until it becomes extremely sticky. The print is then placed face up on the cloth or mat by sticking one edge, and rubbing from this edge toward the opposite side as the print comes in contact with the cloth. This prevents the formation of air bubbles.
3. After print is smooth, roll heavily with a large print roller.
4. The excess rubber cement along the edges is easily rubbed off with a clean, dry towel, after it has completely set.

Making Corrections on Maps.

Maps inked on paper, tracing paper, or tracing cloth may be corrected by erasing either by hand or by the use of an electric eraser. Erasing is a delicate operation, and, to be successful, must be done with care.

1. With a smooth, sharp knife pick the ink from the paper. This can be done without marring the surface.
2. Place a hard, smooth surface, such as a triangle, under the erasure before rubbing starts.
3. When practically all the ink has been removed with the knife, rub with a pencil eraser.

This method requires more patience and skill than the ordinary method of rubbing with an ink eraser until the line disappears, but will leave the surface in much better condition for drawing ink lines. It is

...ntial that the knife be very sharp and have a smooth edge; never attempt this method with a dull knife.

Blue Prints.

Corrections on blue prints are best made with water colors, red and yellow being the colors most often used. Mix the color as thick as it will flow readily from the pen, and rule with right-line pen.

Blue prints are sometimes corrected by using Chinese white or some alkaline solution to produce a white line. This is not as satisfactory as colors, since the original print is also shown in white lines.

Van Dykes are corrected by using red or yellow water colors. Black- or brown-line prints having a white background are the most easily corrected.

1. Draw the changes and corrections in india ink, and allow to dry.
2. Make a bleaching solution as follows: Dilute a small amount of tincture of iodine with an equal quantity of water. Add enough 10 per cent solution of potassium cyanide to the iodine to produce a colorless liquid.
3. With this solution paint out the lines of the print which are to be removed. This solution will not remove or injure the india ink, and will bleach any photographic paper sensitized with a silver solution.
4. The bleaching process should be carried on near a window or in a draft, as the fumes of the cyanide are poisonous.

Black-line prints of topographic maps may be used as the basis of study in drainage, landscaping, etc., the entire results of the study being inked, and all original lines on the print removed by immersion in the above solution. This provides a rapid and accurate method of preparing plans based on the original topography.

Various Devices of Office Practice.

In order to avoid expensive delays in construction, maps are often finished sharp and black in pencil and the penciled drawing given a light brush coat of rapid-drying varnish or shellac. This emphasizes the pencil lines, protects the paper from dirt and moisture, and preserves it in condition for future tracing.

Pencil drawings sprayed with a patent solution called "fixatif" will also accomplish the same results.

Blue prints may be made from typewritten sheets by placing a carbon paper, with the carbon side facing the back of the sheet, directly under the sheet being typed. This gives a double impression of the letter black enough to print by allowing a little extra exposure. Prints may also be made from heavy white paper by soaking in benzine and printing while wet. The benzine will evaporate and leave no trace.

Pencil drawings on paper may also be reproduced by the Photostat method, as described under Reproduction.

Extremely large pencil drawings should be stretched before laying out. This will save the draftsman much time and inconvenience, as most paper is very sensitive to atmospheric changes.

Aids in Lettering.

To the average draftsman, lettering is a major problem. Many mechanical aids have been developed, of which only a few have real merit.

Fig. 113.

Lettering rules similar to Fig. 113 may be purchased at any engineering-supply store. They resemble an ordinary flat scale with the center portion stamped out in the outlines of all the curved-stroke or difficult letters. They are valuable in that they furnish the draftsman a pencil copy in true form.

Many special triangles have been developed as useful aids in lettering. The most commonly used are the Braddock triangles shown in Figs. 114 and 115 and the 70-deg. lettering triangle shown in Fig. 116.

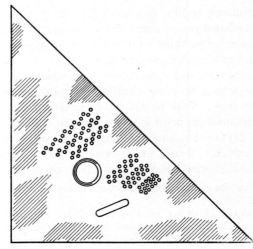

Fig. 114.

Rubber Type Letters can be secured either assembled into a stock title, or as individual letters in alphabets. These letters can be rapidly set up in a printing stock, inked on an ordinary desk pad, and transferred to the drawing. They furnish inked outlines of the letters, correctly spaced, and ready to be traced in

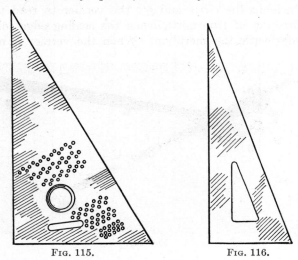

Fig. 115.   Fig. 116.

india ink. Since the ink impression is already made, the india ink flows readily to the same outline.

Printed Stock Titles are useful as time savers. The titles may be printed in sharp clear Roman type at a print shop, placed under the tracing, and traced in India ink.

Mechanical Lettering Devices are being widely used in modern drafting rooms. They are based on the principle of a stylographic pen guided by a master sliding plate and can be purchased on the market under the names of Wrico, Normograph, Leroy, etc.

Care of Field Notes.

Field notebooks, from which the map is plotted, and penciled plane-table sheets constitute the original record of the survey and map. They have an authority in court that no tracing or copy can ever have, and should be filed and preserved in their original form.

Some offices go so far as to require that all notes shall be copied in ink and the original notes indexed and permanently filed in a vault, to guard against loss by fire.

The Universal Drafting Machine.

The universal drafting machine is a labor-saving device that is as much of a boon to the map draftsman as it is to the draftsman engaged in architectural or mechanical work.

The machine consists essentially of two straight-edges or scales mounted on a metal circular plate in such a position that they make an angle of 90 deg. one with the other.

The circular plate is graduated from 0 to 360 deg. in both directions and is fitted with a vernier and clamp similar to the surveyor's transit. Two pairs of double toggle arms connect the circle to a clamp fastened onto the edge of the drawing table.

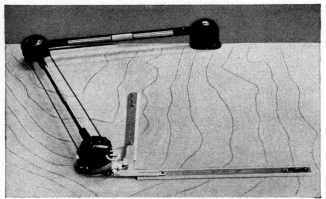

Fig. 116a.—Universal drafting machine.

The toggle arms are so connected that when the straightedges are clamped in any one position, a movement of the machine will always move the straightedges parallel to their fixed position.

The machine may be used to plot directions and distances as used in survey notes as follows: Remove the two straightedges and replace one of them with a scale graduated to the scale of the desired map. Unclamp the circle and set the vernier to read the direction of true north when the scaling edge coincides with the meridian. When the vernier is not

Fig. 117.

clamped the operator can turn the scale to the given azimuth, move the machine until the zero of the plotting scale is over the plotting station, and scale the distance to the required point.

This machine is also a valuable asset in plotting streets, sidewalks, buildings, etc., that are either parallel or perpendicular to each other.

The Polar Planimeter.

The polar planimeter is an instrument used for measuring areas in units of wheel revolutions. Planimeters may have the tracing arm fixed at a constant length or may be fitted with a clamp and slow-motion screw to allow the length of the tracing arm to be adjusted to any desired setting. Figure 117 shows a polar planimeter with an adjustable tracing arm. It consists essentially of four parts: (1) a fixed- or adjustable-length pole arm, (2) an adjustable-length tracing arm, (3) a pivot joint connecting the two arms, and (4) a wheel, disk, and worm-gear assembly that may be clamped to the tracing arm at any desired position. The planimeter is supported on the paper at three points, the anchor point A, the wheel W, and the tracing point T. The adjustable arm L is graduated throughout its effective length. One of these graduations set to the index I gives known relations between wheel revolutions and area traversed. The drum face of wheel W is divided into 100 equal spaces and fitted with a vernier to read fraction parts. Wheel W turns the disk D by means of a worm gear in the ratio of 10 to 1.

Complete revolutions of the wheel are read on disk D, the hundredths of a revolution on the drum of the wheel W, and the thousandths of a revolution on the vernier matching the graduations on the drum. In operating a planimeter it is good practice to start with an initial reading, since it is more accurate than trying to set the vernier of the wheel at zero. The wheel revolutions will then be the difference between the initial and final readings.

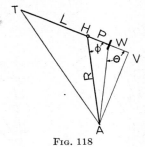

Fig. 118

Theory of Planimeter.[1]—Mathematical proofs of the theory of the polar planimeter by means of the calculus are given in various publications. A direct and simple geometric demonstration of the theory is as follows:

In Fig. 118 the heavy solid lines represent a polar planimeter with the tracing point T, wheel W, and anchor point or pole A. The tracing arm TW is

---

[1] From Davis and Foote, "Surveying," by permission of the authors.

hinged at H to the anchor arm AH. The length of the portion TH of the tracing arm is designated as L and the length of the anchor arm AH as R. (If, as in some designs, the wheel arm HW is folded back on the tracing arm, the relations herein demonstrated still apply.)

from 0 to T, and because the wheel does not rotate while the tracing point is moved along the zero circle from o to 0. The heavy dash lines represent the planimeter with tracing point at t. As the tracing point moves from T to t the wheel moves from W to w, partly by rolling and partly by sliding. The rolling

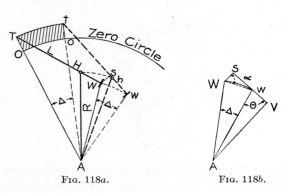

Fig. 118a.   Fig. 118b.

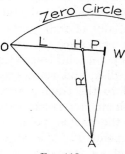

Fig. 118c.

In Fig. 118a consider the infinitesimal area $Tto0$, a portion of the sector TAt, which lies just outside the zero circle. If the tracing point is caused to traverse the perimeter clockwise, the only permanently recorded rotation of the wheel will be that due to the motion from T to t. This is because the movement from t to o is offset by the reverse movement

component of this motion is represented by the line Ws and the sliding component by the line sw.

It will now be shown that the area $Tto0$ is in direct proportion to the roll of the wheel and is therefore equal to a constant K times the number of revolutions n. Reasons for the steps involved in equations (1) to (10) are given immediately following the equations.

Area $Tto0 = \dfrac{\overline{AT \cdot Tt}}{2} - \dfrac{\overline{A0 \cdot 0o}}{2}$ \hfill (1)

Area $Tto0 = \dfrac{\overline{AT^2} \cdot \Delta}{2} - \dfrac{\overline{A0^2} \cdot \Delta}{2}$ \hfill (2)

$ = \dfrac{\Delta}{2}(L^2 + 2LR\cos\phi + R^2 - 2PL - L^2$
$\phantom{Area Tto0 = \dfrac{\Delta}{2}(L^2 + 2LR\cos\phi + R^2 } - R^2)$ \hfill (3)

$ = L \cdot \Delta \cdot (R\cos\phi - P)$ \hfill (4)

$ = L \cdot \Delta \cdot \overline{WV}$ \hfill (5)

$ = L \cdot \Delta \cdot \overline{AW} \cdot \cos\theta$ \hfill (6)

$ = L \cdot \overline{Ww} \cdot \cos\alpha$ \hfill (7)

$ = L \cdot \overline{Ws}$ \hfill (8)

$ = L \cdot c \cdot n$ \hfill (9)

$ = K \cdot n$ where $K = Lc$ \hfill (10)

Equation (1): The area of a sector is one-half of the product of the radius and the subtended arc.

Equation (2): An arc, in linear units, is equal to the product of the radius and the subtended arc in radians.

Equation (3): The lines AT and $A0$ can be expressed in terms of the dimensions of the planimeter, as follows:

In Fig. 118

$\overline{AT^2} = \overline{TV^2} + \overline{AV^2}$
$\phantom{\overline{AT^2}} = (L + R\cos\phi)^2 + (R\sin\phi)^2$
$\phantom{\overline{AT^2}} = L^2 + 2LR\cos\phi + R^2(\cos^2\phi + \sin^2\phi)$
$\phantom{\overline{AT^2}} = L^2 + 2LR\cos\phi + R^2$

In Fig. 118c
$\overline{A0^2} = \overline{0W^2} + \overline{AW^2}$
$\phantom{\overline{A0^2}} = (P + L)^2 + R^2 - P^2$
$\phantom{\overline{A0^2}} = 2PL + L^2 + R^2$

Equation (4): Collect terms of equation (3).

Equation (5): In Fig. 118

$\overline{WV} = \overline{HV} - P = R\cos\phi - P.$

Equation (6): From geometry of Fig. 118.

Equation (7): An arc, in linear units, is equal to the product of its radius and the subtended angle in radians, hence in Fig. 118b the arc

$\overline{Ww} = \Delta \cdot \overline{AW} = \Delta \cdot \overline{Aw}.$

Line Ws is perpendicular to line SV; in the limit, line Ww is arc Ww and is perpendicular to line Aw; hence $\cos \cdot \alpha = \cos\theta$.

Equation (8): From geometry of Fig. 118b.

Equation (9): The roll of the wheel is the product of the length of its circumference c and the number of revolutions n.

Equation (10): The two characteristic dimensions of the instrument L and c are grouped into one instrument constant K.

For a figure drawn at natural scale, the planimeter constant K is equal to the product of the length of the tracing arm and the circumference of the wheel. For figures drawn at other scales, the planimeter constant may include the factor for converting actual areas to areas at the given scale; in this case the constant is usually designated as K'.

To Determine Planimeter Constant K.—Lay out a square or rectangle very carefully and determine its area in square inches by computation. Calculate the length of the diagonals and check by scaling for length and equality. Clamp the tracer arm at a convenient working length, set the anchor point outside the square or rectangle, and traverse the area in a clockwise direction. The difference between the initial and final readings will be the area in wheel units. Repeat the operation three times and take the mean as the number of wheel units.

The constant K is determined by dividing the known area of the square or rectangle by the average determination of wheel revolutions. The wheel reading for any other area multiplied by K will give the area traversed in square inches. When the scale of the map is known, the area in square inches may be converted to any other units of area.

Example.—A 4- by 4-in. square is traversed with a planimeter three times. The average of the three differences is 1.542. On a map whose scale is

$$1 \text{ in.} = 100 \text{ ft.}$$

a traversed area reads 2.125 in wheel revolutions. What is the area in acres?

Solution.

$$K = \frac{4 \times 4}{1.542} = 10.376$$

$$2.125 \times 10.376 = 22.049 \text{ sq. in. map area}$$

$$\frac{22.049 \times 100 \times 100}{43,560} = 5.06 \text{ acres}$$

To Adjust Planimeter to Read in Map Units.—The tracer arm on the planimeter can be set so that the wheel will read area in any units wanted on a map of any scale by making use of equation (9),

$$A = L \cdot c \cdot n$$

in which L = length of tracer arm, inches.
c = circumference of wheel, inches.
n = number of revolutions of the wheel.
A = number of square inches in unit of area wanted based on the scale of the map.

In order to use $A = L \cdot c \cdot n$ it is first necessary to determine the value of c as follows:

1. Lay out a square of known area.
2. Clamp tracer arm in position.
3. Measure tracer arm from pivot point to tracing point.
4. Traverse square three times and take the mean of wheel revolutions.
5. $c = \dfrac{A}{L \cdot n}$, where A, L, and n are known.

Example.—Use a 6- by 6-in. square, $A = 36$ sq. in.
$L = 5.61$ in. by measurement.
2.667 = mean value of wheel revolutions for three determinations.
Then

$$c = \frac{36}{2.667 \times 5.61} = 2.406 \text{ in.} = \text{circumference of wheel.}$$

Having determined the value of c, suppose it is required to find the setting of the tracer arm L that will give wheel readings in square miles on a map whose scale is 1 in. = 1,000 ft. or 1/12,000.

Map units are as follows:

1 sq. in. = 1,000,000 sq. ft.
1 sq. mi. = 27,878,400 sq. ft.

$$A = \frac{27{,}878{,}400}{1{,}000{,}000} = 27.8784 \text{ sq. in. on map} = 1 \text{ sq. mi.}$$

As the wheel must give the area in square miles, one revolution of the wheel = 1 sq. mi. of area and $n = 1$.
From $A = L \cdot c \cdot n$

$$L = \frac{A}{n \cdot c} = \frac{27.8784}{1 \times 2.406} = 11.58 \text{ in.}$$

The distance 11.58 in. is too great to be set on the tracer arm. If both sides of the equation $L = A/n \cdot c$ be divided by two the relationship will remain the same. If L is divided by two, n must also be divided by two. Thus an arm setting of 5.79 in. may be used and the final result divided by 2.

The same results are arrived at by the use of the representative fraction scale directly.

$$A = \frac{5{,}280 \times 5{,}280 \times 144}{12{,}000 \times 12{,}000}$$
$$n = 1$$
$$c = 2.406$$

Then $\quad L = \dfrac{5{,}280 \times 5{,}280 \times 144}{12{,}000 \times 12{,}000 \times 1 \times 2.406} = 11.58$ in.

Precautions in the Use of a Planimeter.—If the anchor point is placed outside the area to be measured and the tracing point moved in a clockwise direction around the area, the differences between initial and final readings will always be positive and no confusion exists. If the anchor point or pole is placed inside the area to be measured the determination of the correct area is not so simple. At a certain fixed distance between the anchor point and the tracing point the wheel will not rotate but will slide on the map. Held in this fixed relation, the tracing point may travel completely around the anchor point on the circumference of a circle (known as the zero circle) without any recorded motion of the wheel. When the anchor point is inside the area several important facts must be recognized.

1. The planimeter does not record the actual area traversed but measures the difference between that area and the area of the zero circle.

2. If the final reading is greater than the initial, the area is greater than the area of the zero circle, and, conversely, if the final reading is less than the initial the area is less than that of the zero circle.

3. For clockwise rotation the net travel of the wheel will be forward for area greater than the zero circle and backward for areas less than the zero circle.

4. It follows that for areas known to be less than the zero circle the differences will be positive if the traversing is done *counterclockwise*.

To Determine the Area of the Zero Circle.—The area of the zero circle can be determined by placing the anchor point outside a rectangle of known area and traversing it clockwise. Then place the anchor point near the center of the area and traverse clockwise. The area of the zero circle in wheel revolutions is equal to the difference of the two differences, taking due note of positive and negative readings.

Example 1.—A planimeter has a constant of 10.376. Traversed with anchor point outside the area difference equals 3.469. With anchor point inside the area, difference equals minus 8.641.

$3.469 - (-8.641) = 12.11$ area zero circle in wheel revolutions

$12.11 \times 10.376 = 125.65$ sq. in. area zero circle

Example 2.—Using the same setting on the planimeter as in Example 1, an area judged to be larger than the area of the zero circle was traversed clockwise with the anchor point inside the area. Initial reading was set at .011 and the final reading 5.631.

$(5.631 - .011) \times 10.376 = 58.31$ sq. in. difference between traversed area and area of zero circle

$58.31 + 125.65 = 183.96$ sq. in. traversed area

Example 3.—On an area smaller than the area of the zero circle. Anchor point inside area and traversed clockwise. Initial setting 9.962. Final reading 6.742, difference is minus 3.22 revolutions.

3.22 × 10.376 = 33.41 sq. in. area is less than area of zero circle

125.65 − 33.41 = 92.24 sq. in. of traversed area

Some operators prefer to set the initial reading at some low figure, say, 1.123, and traverse the area *counterclockwise*. This would have given a final positive reading of 4.343. The difference would still be 3.22 revolutions and would be reckoned minus since it was traversed counterclockwise.

Accuracy of Planimeter Measurements.—The precision of planimeter measurements depends upon several factors:

1. The care and accuracy with which the planimeter constant is determined.
2. Carefulness in setting the length of tracer arm if it is desired to read areas in terms of map units.
3. Skill in following the boundary line of the area with the tracer point.
4. Scale of the plotted areas to be traversed.

Too much emphasis cannot be placed upon the determination of the constant.

The trial area (square or rectangle) should be accurately plotted and rechecked by scaling. The diagonals should be calculated and then carefully scaled to see that they are the same length and scale the calculated distance. The operator should secure three to five readings on the trial area that agree very closely. The mean of the above readings is taken as the correct value for the number of wheel revolutions.

It is also very difficult to set the tracer arm at a calculated length and get an exact setting the first time. The adjustable arm is fitted with a slow-motion screw for movement of the arm in small increments in either direction. Trials must be repeated until the wheel reading will agree with the known area traversed.

Lack of skill in moving the tracing point along the boundary line is one of the main sources of error in planimeter work. Since the error is as liable to be on one side of the boundary line as on the other, these errors are largely compensating; however, it should be apparent to the operator that these errors are a much larger percentage of a small area than they are of a large area. It follows that the greater the accuracy desired the larger the scale to which the areas should be plotted. With the occasional operator of a planimeter it is possible to keep the error under 2 per cent of the area measured even on small areas.

On larger areas the same operator may reduce the error to one-eighth of this amount.

Other Types of Planimeters.—Planimeters with which the engineer and draftsman are most familiar are of the polar planimeter type with either fixed or adjustable arms. Their function is to determine areas with one simple operation. Special types of planimeters are manufactured to perform specific functions, and the student may gain some familiarity with them by consulting the catalogues of various instrument makers. *Radial* planimeters are designed for the purpose of measuring mean heights of circular diagrams with uniformly spaced ordinates. Where the circular ordinates are spaced with any other than the first power of the quantities that they represent, a special radial planimeter must be designed for this particular chart and made to order.

The *rolling planimeter* moves on two broad rollers, from one of which motion is imparted to the recording mechanism. Since only the rollers and tracer are in contact with the drawing, the results are not affected by any irregularities in the surface of the paper. They are very accurate and have the advantage that a strip of any length and of a width not exceeding the length of the tracer arm can be measured at one operation. The *disk* or *suspended planimeter* is the same in principle as the polar planimeter but has the advantage of allowing the measuring wheel to move over a smooth plate or disk. Both the disk and rolling planimeters are precision instruments and are more accurate than the polar planimeter.

As with the polar planimeter the error of both disk and rolling planimeters varies with the size of the area measured. The disk planimeter will give errors about one-sixth of the error of a polar planimeter and the rolling planimeter about one-tenth of the same error.

A precision planimeter combining the features of both the disk and the rolling type is also on the market and is known as the *rolling-disk planimeter*.

Integrators are built much on the same principle as planimeters but will ascertain the area and moments relative to any axis of any figure by simply tracing its outline.

The integraph is another form of integrator that draws automatically the integral curves. This gives a graphic representation of the integration, a feature very valuable in some lines of engineering work.

Drawing Papers.

The selection of a drawing paper for a given purpose requires the consideration of its body, surface, erasing quality, permanency, color, expansivity, and cost.

Body.—The term "body" as here used applies primarily to the strength of a paper. Any paper should be strong enough to withstand satisfactorily the handling it will get under expected conditions. Working drawings and prints require a paper capable of standing up well under severe handling conditions.

Surface.—The type of surface to be used depends on whether the drawing is to be in ink or color or to remain in pencil. Pencil drawings can be satisfactorily made on paper of average roughness, while ink drawings generally require a smooth surface, particularly if made for reproduction. Color drawings usually require a rough sand-grained or pebbled surface.

Erasing Quality.—The measure of a paper's erasing quality is its ability to withstand a number of erasures of both pencil and ink lines over the same spot and still have a surface at that particular spot that will permit the drawing of sharp, clean-cut ink lines.

Permanency.—The permanency of a paper should be such as to ensure a form and color that do not change within its required life. The best grades of drawing papers are made from rags that have not undergone wear. Vegetable fibers are also used, but fine paper depends largely on high-grade rags for its qualities of permanence and strength.

Color.—Fine drawings require a clear white paper, while working drawings may be put on tinted or colored paper that will not show dirt so easily. Cream-colored papers may be used for both purposes.

Expansivity.—Paper expands and contracts with variation in the humidity of the atmosphere, and the effects of such changes on the scale of a drawing are very important. This is particularly true when the expansion or contraction is not the same amount in all directions, resulting in a change not only in scale but in the form of the drawing. Distortion produced by expansion or contraction can be alleviated somewhat by tacking the paper or tracing cloth to a board or wall and letting it season or air-condition before use. Another aid in reducing distortion due to expansivity is to lay out the long dimension of the drawing in the machine direction rather than the cross direction of the paper.

Clear-cut conclusions and comprehensive data based on actual tests of various commonly used drawing papers are difficult to obtain, but based on the results of isolated tests certain characteristics stand out as fairly well established. One such characteristic is that shrinkage and expansion in both drawing papers and tracing cloths is much greater in the cross direction than in the machine direction.

Another clear-cut characteristic that seems well established is that in the case of tracing cloth the shrinkage is particularly severe immediately after the cloth is taken from the roll, before it has an opportunity to air-condition.

In tests made on various papers it has been found that the normal bond papers change in dimension about 0.6 per cent in the machine direction and 1.6 per cent in the cross direction when the relative humidity varies from 95 to 15 per cent. This is regardless of base weight or rag content.[1]

Cost.—In many instances the results of many weeks or months of surveys are shown in final assembled form on a drawing. Such being the case, the quality of the paper or tracing cloth used should be the best obtainable for the purpose regardless of cost.

Types of Paper.

For fine drawings in ink, bristol board of good quality is recommended, such as Reynold's patent office bristol board, which comes in 2-, 3-, and 4-ply sheets in sizes 8 by 13 in., 10 by 15 in., 15 by 20 in.,

[1] Bralkowsky and Probst, *Paper Trade Journal*.

20 by 30 in., etc., or Strathmore 2- and 3-ply bristol board in sheets 8 by 13 in., 10 by 15 in., etc. Whatman's hot-pressed drawing paper, which comes in sheets 13 by 17 in., 15 by 20 in., 17 by 22 in., 19 by 24 in., 22 by 30 in., 27 by 40 in., 31 by 53 in., is also an excellent paper for pen-and-ink work.

For water-color work Whatman's cold-pressed (grained) paper, which comes in sheets 13 by 17 in., 15 by 20 in., 17 by 22 in., 19 by 24 in., 22 by 30 in., 27 by 40 in., 31 by 53 in., etc., or paper of similar quality should be used.

Whatman's rolled paper in sheets 19 by 24 in., 22 by 30 in., 27 by 40 in., etc., is also good for color work.

Detail paper of good quality, either white or cream in color, may be used for certain types of mapping, particularly where the map is to be traced. Such paper usually comes in rolls 30, 36, 42, 54, 58, 62, 72 in., etc., wide, the rolls containing from 10 to 50 yd. of paper.

For drawings subjected to hard wear or great changes in humidity, papers similar to the Paragon papers mounted on muslin may be used. These papers will usually come in sheets 15 by 15 in., 18 by 24 in., 24 by 31 in., etc., or in 10-yd. rolls

36, 42, 62 in., etc., wide. Tracing cloth to give satisfactory results should be of a quality equal to Imperial tracing cloth. This cloth comes in 24-yd. rolls 24, 30, 36, 38, 42, 48, 54 in., etc., wide.

Where changes in humidity or temperature constitute an important factor, sheets of cellulose acetate may be used to advantage as it expands and contracts at about the same rate in all directions, hence eliminates distortion and reduces change in scale to a minimum. The coefficient of linear expansion of cellulose acetate used by the U. S. Geological Survey is about 0.0009 per degree Fahrenheit.

Pens.

It is difficult to be arbitrary in the matter of lettering pens owing to differences in the pressure applied on a pen by different draftsmen in lettering.

Depending on the pressure applied, good results may be obtained by any of the following pens or pens of similar quality: Gillott 303, 404, Hunt 512, Leonardt 516F, 506F, Esterbrook 1000, 968, Spencerian No. 1, etc.

For mapping pens, where fine lines are required, Gillott 290, 291, 170 and Esterbrook 356, 355, or the equivalent should give good results.

CHAPTER IX

SUGGESTED PROBLEMS

In order that the student may become proficient in both drawing and interpreting contour forms, it is necessary to present him with many varied and irregular problems in contour sketching. These problems naturally divide themselves into the following forms:

1. Picturing natural formations by means of contours.
2. Construction of profiles from contour maps.
3. Problems in contour crossings.
4. Correct interpretation of relief features.
5. Picturing changes in relief due to grading roads, railroads, levees, etc.
6. Problems which may be solved from contour maps.

Picturing Natural Formations.

Select a portion of terrain of which the instructor has an accurate contour map. Furnish the student with a few control elevations and require him to sketch a plan view by means of contours and a right section through a portion of the plan view. When finished, compare with the original map and point out his mistakes in interpretation. Figure 119 illustrates in brief form the type of problem used. Section AB represents a saddle between two hills, section CD a shallow valley between two ridges, section EF an elongated mound 40 ft. high, section GH an overhanging cliff, section KL a roughly circular depression 40 ft. deep, and section MN the location of a winding contour from interpolation. After the student has gained a certain degree of proficiency in contour sketching of natural terrain, the following type problems may be assigned as classroom work:

1. Represent by contours the crater of an extinct volcano roughly circular in shape. The top diameter is 200 ft. and the bottom diameter 50 ft., depth, 60 ft.; top width of crater wall approximately 12 ft.; and outside slopes roughly 2 to 1. Slopes are cut by small, irregular gullies.

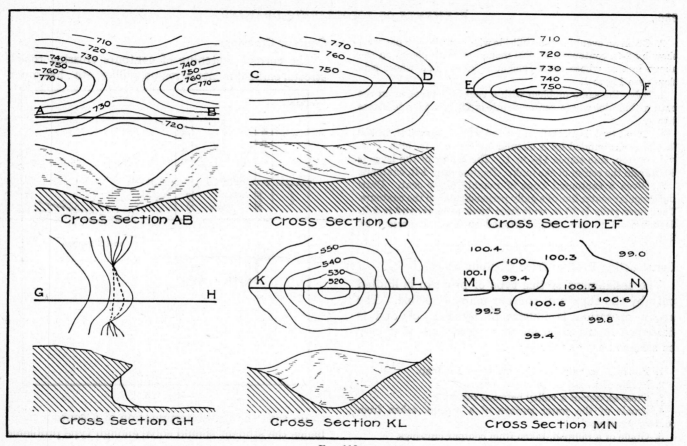

Fig. 119.

2. Given an evenly sloping hillside whose grade is 10 per cent, show by 2-ft. contours a conical mound of earth placed on the above hillside whose center height is 23 ft., and whose side slopes are 1½ to 1.

3. Solve the same problem with a conical depression on the natural hillside, whose center depth is 23 ft. and side slopes 1 to 1.

4. An oil tank is located on ground which slopes 3 ft. in 100 ft. It is surrounded by a circular fire levee 5 ft. high, 6 ft. wide on top, side slopes 1½ to 1. The center of the levee is 80 ft. from the center of the tank. Represent by 1-ft. contours.

5. Given a contoured slope of roughly 6 per cent. Draw the irregular center line of a gully perpendicular to the contours. The above gully is 10 ft. wide from bank to bank, and is 5 ft. deep normal to the natural ground surface. Represent by 2-ft. contours.

Constructing Profiles from Contour Maps.

The assignment of this type of problem is limited only by the supply of contour maps available. The above operation is illustrated by Fig. 57 and is described in the adjacent text. Problems that may be assigned are as follows:

1. Profile of any straight line drawn on the map.
2. Profile of a road between two towns.
3. Cross-section of a valley between two hills.
4. Profile of the water surface of a stream, taken from contour crossings.
5. Grade of a railroad, taken from contour crossings.

Contour Crossings.

The correct sketching of contours crossing roads, streets, embankments, streams, gullies, etc., has much

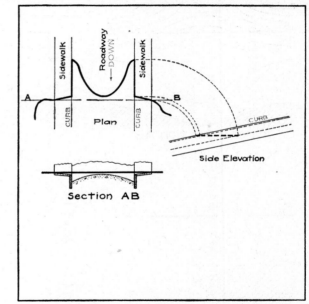

Fig. 120.

to do with the final interpretation of a contour map, hence the student should solve enough type problems to become familiar with this part of contour sketching. Figure 120 shows a contour crossing a paved street, while Fig. 121 shows a contour crossing a railroad

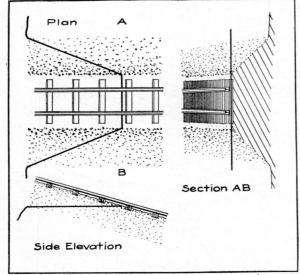

Fig. 121.

embankment. By reference to certain facts concerning contours listed in the chapter on Contours and Contour Sketching, this type of problem is easily solved. Problems 1 to 4 in Fig. 123 are illustrative of this type, and are self-explanatory.

Interpreting Relief Features.

Figure 122 represents an area where a portion of the contours have been incorrectly sketched to illustrate common errors:

A. Contours cannot cross, except in case of an overhanging cliff, and then the portion of the 136 contour looping the 138 should be dotted.

B. Represents an incorrectly sketched saddle.

C. Shows the contours correctly sketched with reference to the ridge line.

D. Shows a ridge line whose contour crossings are incorrect. (It should be remembered that contours never split or merge into one, as shown at *I.*)

E. Shows a stream system whose contours are sketched to fit the drainage lines, while on the system shown at

F. The contour crossings are incorrect.

G. Represents a saddle, correctly sketched.

H. Represents the common error of dropping a contour without its either closing on itself or running to the limits of the map.

Problems in Interpretation.

1. Trace the contours from a topographic map, taking care to portray stream crossings accurately. Omit all streams and gullies. Have the student interpret streams, gullies, and ridge lines from the contours, and compare with the original map.

Fig. 122.—Mistakes in contour interpretation.

2. Trace all streams, gullies, and ridge lines from a portion of a contour map. Supply interpolated elevations along the stream and ridge lines, as shown in Fig. 60. Also, give elevations and location for all saddles and tops of hills. Have the student plot the contours from the above information, and compare with original map.

3. Lay out a system of squares as shown in Fig. 61, and record elevations for all intersections. Have the student plot the contours and determine ridge and drainage lines.

Problems from Topographic Sheets.

The U. S. Geological Survey quadrangle sheets may be purchased from the director, U. S. Geological Survey, Washington, D. C. Have each member of the class supply himself with one quadrangle sheet, from which the following type problems may be assigned:

1. Construct a profile of a road between two points. Method shown in Fig. 57 and described in text.

2. Draw an east and west line on the map and have the student construct an elevation looking north. (Method shown and described under Fig. 62.)

3. Draw a line at right angles to a stream, representing a proposed dam site. Have the student determine the drainage area and flooded area of a dam of given elevation, as shown and described under Fig. 59.

4. Superimpose the center line of a road location, either straight or curved, on the contour map. Supply the student with a proposed grading section, starting elevation, and per cent of grade. Have the student draw the contours as they will appear when grading is finished. Refer to discussion under Fig. 63.

5. Have the student locate a highway or railroad between two points on the map and plot a profile, grade line, and alignment diagram, as shown and discussed under Fig. 64.

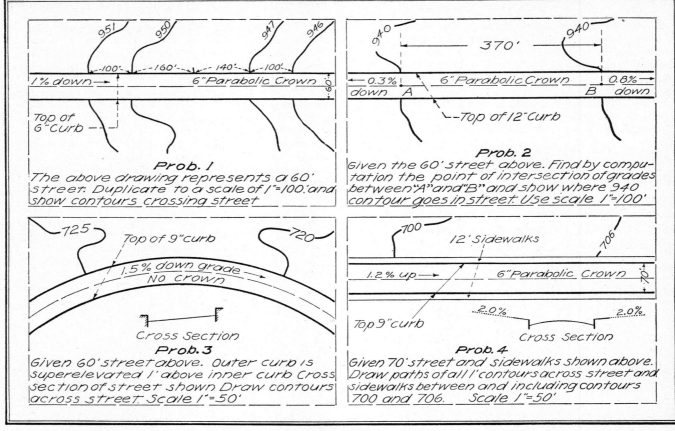

Fig. 123.

FEDERAL GOVERNMENT ORGANIZATIONS USING OR PRODUCING MAPS AND IN SOME CASES ORIGINATORS OF CONVENTIONAL SIGNS, SYMBOLS, AND ABBREVIATIONS

Aeronautics Branch, Department of Commerce.
Air Service, War Department.
Biological Survey, Department of Agriculture.
Bureau of Aeronautics, Navy Department.
Bureau of Agricultural Engineering, Department of Agriculture.
Bureau of the Census, Department of Commerce.
Bureau of Insular Affairs, War Department.
Bureau of Public Roads, Department of Agriculture.
Bureau of Reclamation, Department of the Interior.
Bureau of Soils, Department of Agriculture.
Bureau of Yards and Docks, Navy Department.
Corps of Engineers, War Department.
Division of Maps, Library of Congress.
Division of Topography, Post Office Department.
Federal Power Commission.

Forest Service, Department of Agriculture.
General Land Office, Department of the Interior.
General Staff, War Department.
International Boundary Commission, Department of State.
Mississippi River Commission, War Department.
National Park Service, Department of the Interior.
Office of the Geographer, Department of State.
Office of Indian Affairs, Department of the Interior.
Quartermaster General's Office, War Department.
U. S. Coast and Geodetic Survey, Department of Commerce.
U. S. Geological Survey, Department of the Interior.
U. S. Hydrographic Office, Navy Department.
U. S. Lake Survey, War Department.
Weather Bureau, Department of Agriculture.

REFERENCES

DRAWING AND LETTERING

Ames, Irene K.: "A Portfolio of Alphabet Designs," John Wiley & Sons, Inc., New York, 1938.

Carter, I. N.: "Engineering Drawing," International Textbook Company, Scranton, Pa., 1939.

DeGarmo, E. P., and F. Jonassen: "Technical Lettering," The Macmillan Company, New York, 1941.

French, T. E.: "Engineering Drawing," 6th ed., McGraw-Hill Book Company, Inc., New York, 1942.

———, and Robert Meikeljohn: "Essentials of Lettering," McGraw-Hill Book Company, Inc., New York, 1912.

———, and W. D. Turnbull: "Lessons in Lettering," Books I and II, McGraw-Hill Book Company, Inc., New York, 1942.

Giesecke, F. E., A. Mitchell, and H. C. Spencer: "Technical Drawing," The Macmillan Company, New York, 1936.

Reinhardt, C. W.: "Lettering for Draftsmen, Engineers and Students," 14th ed., D. Van Nostrand Company, Inc., New York.

Rollinson, Charles: "Alphabets and Other Material Useful to Letterers," 1st ed., D. Van Nostrand Company, Inc., New York.

Schumann, C. H.: "Lettering Practice Layouts," D. Van Nostrand Company, Inc., New York, 1936.

———: "Technical Drafting," Harper & Brothers, New York, 1940.

Svenson, Carl L.: "The Art of Lettering," D. Van Nostrand Company, Inc., New York.

———: "Drafting for Engineers," 2d ed., D. Van Nostrand Company, Inc., New York, 1939.

———: "Essentials of Drafting," 3d ed., D. Van Nostrand Company, Inc., New York, 1943.

GEOLOGY AND GEOMORPHOLOGY

Gregory, H. E., et al.: "Military Geology and Topography," Yale University Press, New Haven, 1918.

Emmons, W. H., G. A. Thiel, C. R. Stauffer, and I. S. Allison: "Geology," McGraw-Hill Book Company, Inc., New York, 1932.

Hinds, N. E. A.: "Geomorphology," Prentice-Hall, Inc., New York, 1943.

Legget, R. F.: "Geology and Engineering," McGraw-Hill Book Company, Inc., New York, 1939.

Lobeck, A. K.: "Geomorphology," McGraw-Hill Book Company, Inc., New York, 1939.

———: "Block Diagrams and Other Graphic Methods Used in Geology and Geography," John Wiley & Sons, Inc., New York, 1924.

Longwell, Chester, Adolf Knopf, and R. F. Flint: "Textbook of Geology," 2d ed., Part 1, John Wiley & Sons, Inc., New York, 1939.

Miller, William J.: "Elements of Geology," 2d ed., D. Van Nostrand Company, Inc., New York, 1939.

———, "Introduction to Physical Geology," 4th ed., D. Van Nostrand Company, Inc., New York, 1941.

Ries, Heinrich, and T. L. Watson: "Engineering Geology," John Wiley & Sons, Inc., New York.

von Engeln, O. D.: "Geomorphology Systematic and Regional," The Macmillan Company, New York.

Worcester, "A Textbook of Geomorphology," D. Van Nostrand Company, Inc., New York, 1939.

Numerous publications of the U. S. Geological Survey, U. S. Department of Interior, Government Printing Office, Washington, D. C.

Publications of The American Geographical Society, New York.

MAP PROJECTIONS AND GEODESY

Adams, O. S., and C. N. Claire: "Manual of Plane Co-ordinate Computation," Special Publication 193, U. S. Coast and Geodetic Survey, Government Printing Office, Washington, D. C., 1935.

———: "Manual of Traverse Computation on the Lambert Grid," Special Publication 194, U. S. Coast and Geodetic Survey, Government Printing Office, Washington, D. C., 1935.

———: "Manual of Traverse Computation on the Transverse Mercator Grid," Special Publication 195, U. S. Coast and Geodetic Survey, Government Printing Office, Washington, D. C., 1935.

Birdseye, C. H.: "Formulas and Tables for the Construction of Polyconic Projections," Bulletin 809, U. S. Geological Survey, Government Printing Office, Washington, D. C.

Bowie, W., and O. S. Adams: "Grid System for Progressive Maps in the United States," Special Publication 59, U. S. Coast and Geodetic Survey, Government Printing Office, Washington, D. C., 1919.

Cary, Edward R.: "Geodetic Surveying," John Wiley & Sons, Inc., New York, 1916.

Crandall, C. L.: "Textbook on Geodesy and Least Squares," John Wiley & Sons, Inc., New York, 1907.

Deetz, Charles H., and Oscar Adams: "Elements of Map Projection with Applications to Map and Chart Construction," Special Publication 68, U. S. Coast and Geodetic Survey, Government Printing Office, Washington, D. C., 1928.

Hinks, Arthur R.: "Map Projections," 2d ed., The Macmillan Company, New York, 1921.

Hosmer, O. L.: "Geodesy," 2d ed., John Wiley & Sons, Inc., New York, 1930.

Melluish, R. K.: "An Introduction to the Mathematics of Map Projections," The Macmillan Company, New York, 1931.

Sharp, H. O.: "Geodetic Control Surveys," 2d ed., John Wiley & Sons, Inc., New York, 1943.

Whitmore, G. D.: "Geodetic Surveying," International Textbook Company, Scranton, Pa., 1942.

Other publications of the U. S. Coast and Geodetic Survey dealing with geodesy, map projections, and related subjects include Special Publications 5, 57, 49, 52, 53, 57, 60, 112, 130, and 153.

MAPS, TOPOGRAPHY, AND PHOTOGRAMMETRY

Bouchard, Harry: "Surveying," 2d ed., International Textbook Company, Scranton, Pa., 1940.

Breed, C. B., and G. L. Hosmer: "Elementary Surveying," 7th ed., John Wiley & Sons, Inc., New York, 1938.

———: "Higher Surveying," 5th ed., John Wiley & Sons, Inc., New York,

Church, Earl: "Mathematical Computations in Aerial Photogrammetry," Syracuse University Press, Syracuse, N. Y., 1936.

Dake, C. L., and J. S. Brown: "Interpretation of Topographic and Geologic Maps," McGraw-Hill Book Company, Inc., New York, 1925.

Davis, Raymond E., and Francis S. Foote: "Surveying," 3d ed., McGraw-Hill Book Company, Inc., New York, 1940.

Deetz, Charles H.: "Cartography," U. S. Coast and Geodetic Survey, U. S. Department of Commerce, Washington, D. C., 1936.

Field, R. M., and H. T. Stetson: "Map Reading and Navigation," 1st ed., D. Van Nostrand Company, Inc., New York, 1942.

Field Manual 21-30, "Conventional Signs, Military Symbols and Abbreviations," War Department, Government Printing Office, Washington, D. C., 1941.

Finch, J. K.: "Topographic Maps and Sketch Mapping," John Wiley & Sons, Inc., New York, 1920.

Fordham, Sir Herbert G.: "Maps, Their History, Characteristics and Uses," The Macmillan Company, New York, 1929.

Hinks, Arthur R.: "Maps and Survey," 4th ed., The Macmillan Company, New York, 1942.

"Manual," The American Railway Engineering Association, Chicago, Ill., 1943.

Publications of the American Society of Photogrammetry, Washington, D. C.

Putnam, William C.: "Map Interpretation with Military Applications," McGraw-Hill Book Company, Inc., New York, 1943.

Raisz, Erwin: "General Cartography," McGraw-Hill Book Company, Inc., New York, 1938.

Rayner, W. H.: "Advanced Surveying," D. Van Nostrand Company, Inc., New York, 1941.

Reed, Brigadier General Henry A.: "Topographical Drawing and Sketching," John Wiley & Sons, Inc., New York.

Roberts, L. B.: "Topographic Mapping," The Society of American Military Engineers, Washington, D. C., 1924.

Rubey, Harry, G. E. Lommel, M. W. Todd: "Engineering Surveys," The Macmillan Company, New York, 1942.

Salisbury, R. D., and W. W. Atwood: "The Interpretation of Topographic Maps," Professional Paper 60, U. S. Geological Survey, U. S. Department of Interior, Washington, D. C., 1908.

Sharp, H. Oakley: "Photogrammetry," 3d ed., John Wiley & Sons, Inc., New York, 1943.

Stuart, Edwin R.: "Topographical Drawing," McGraw-Hill Book Company, Inc., New York, 1917.

Talley, Captain B. B.: "Engineering Applications of Aerial and Terrestrial Photogrammetry," Pitman Publishing Corporation, New York, 1938.

Technical Manual TM 5-230, "Topographic Drafting," U. S. War Department, Washington, D. C., 1940.

Training Manual 2180-5, "Map and Aerial Photograph Reading," U. S. War Department, Government Printing Office, Washington, D. C., 1938.

Whitmore, George D.: "Elements of Photogrammetry," International Textbook Company, Scranton, Pa., 1942.

Wilson, H. M.: "Topographic, Trigonometric and Geodetic Surveying," John Wiley & Sons, Inc., New York, 1912.

INDEX

A

Accuracy, tests for, 23
Adjustment of highway surfaces, 169
Aids in lettering, 220
Air-navigation symbols, 79
Airport maps, 16
Alluvial fans, 151
Ames protractor, 104
Authorized abbreviations, XV
Azimuthal map projection, 197

B

Bar scales, 22
Beam compass, 212
Blending of tints, 131
Blocking out details (color topography), 138
  of map, 103
Blue prints, making corrections on, 219
  method of making, 185
Border lines, 120

Braddock triangle, 220
Brushes, testing, 126
  for water coloring, 126
Building site maps, information required on, 4

C

Cadastral maps, 2
Checkerboard map, 160
Checking, map details, 106
City maps, reproduction of, 189
Civil and political divisions, lettering used, 115
Classification of maps, 2, 14, 18
Cliffs, 109
Cloth-mounted maps, 217
Colored pencils, use of, 139
Colors, durability of, 138
  preparation of, 128
  topographic drawing in, 125
  tube or pan, 125
  used on geological maps, 139

Colors, values and durability of, 128, 138
Composition, 26
Conic map projection, 201
Containers for color mixing, 126
Contours, in color topography, 137
  from controlling points, 160
  discussion of, 143
  emphasized, 108
  facts concerning, 143
  faults in drawing, 39
  forms, 148
  map, penciling of, 108
    side elevations, 163
    use of, 154
  plotting from profiles, 160
  sketching, 145
Conventional signs (*see* Symbols)
  in color, 133
Conversion scale, 21
Copying, notes on, 179
Corrections on maps, 218

Crossings, contour, 236
  stream, 109
Culture, definition of, 2
Culture symbols, 52–54, 61–67, 80
Curves, irregular, 214
  railroad, 215
  use of, 214
Cuts and fills, 109
Cycles, erosion, 150
Cylindrical map projection, 199

D

Depression contours, 109, 174
Depressions, 109
Development of symbols, 27
Diagonal scale, 23
Display maps, mounting, 216
Dividers, proportional, 183, 210
Drafting machine, universal, 221
Dragging, 132
Drainage, penciling of, 108
Drainage areas, 157
Drainage lines, 108
Drawing for reproduction, 192
Drawing instruments, care of, 213
Drawing papers, 230–232
Duplication, methods of, 185
  by photostat, 187
  by zinc etchings, 187

E

Earthwork, contour problems in, 239
Engraving and printing U.S.G.S. maps, 190
Erosion, wind, 151
Erosion cycles, 150

F

Federal Government organizations using maps, 241
Field notes, care of, 221
Forest Service symbols, 58
Formations, mountain, 154
Forms, glossary of topographic, 172

G

Gas and oil symbols, 77
Geological maps, colors used on, 139
Geological sections, 84, 85
Geological Survey maps, U.S., (facing), 14, 154, 190
Geological symbols, 81
Geology, relation to topography, 141
Glaciation, 148
Glass-top table, 179
Glossary of topographic forms, 172
Gnomonic map projection, 198

Golf-course maps, information required on, 8
Golf-course symbols, 78
Grid system, military, 205

H

Hachure scale, 43
Hachures, definition, 41
  use of, 109
Highway location plan, 167
Highway plan and profile, 166
Highway surface, adjustment by contours, 169
Hill shading, 40–41, 47
  precautions, 42
  rules for, 42–43
Hydrographic maps, information required on, 8
Hydrographic symbols, 46, 55–57
Hydrography, definition of, 2
  lettering used on, 116
Hypsographic symbols, 39, 51
Hypsography, lettering used on, 116

I

India ink, use in water-color topography, 137
Inking, general rules for reproduction, 193
  order of, 110

Inking, topographic maps, 109
Instruments, care of, 213
    not in common use, 208
Integraph, 230
Integrators, 230
Interpretation, problems in, 237
    of relief features, 145, 237

L

Landscape maps, 11
Lambert conformal map projection, 203
Latitudes and departures, use in plotting traverses, 98
Laying tints, precautions, 131, 137
Lettering, aids in, 220
    alphabets, 121–124
    civil and political discussions, 115
    in color topography, 138
    construction of, 114
    hydrography, 116
    hypsography, 116
    inclination of, 115
    public works, 117
    for reproduction, 193
    rules for size and distribution, 113
    for topographic maps, 111
Limits of topographic expression, 144
Line drawings, in reproduction, 193
Location, paper location from contour maps, 165

M

Map reproduction, 189
Maps, airport, 16
    cadastral or city, 2
    city, reproduction of, 189
    classified, as to delineation, 18
        as to use, 2
    cloth mounted, 217
    colored, 14
    corrections on, 218
    general, 8
    geographic, 2
    golf, polo, 8
    hydrographic, 6
    inking of, 109
    landscape, 11
    lettering, 111
    military, 14
    picture or display, 8
    projections, 195
    reduction or enlargement, 180
    scales, 18
    tests for accuracy, 23
    titles, 117
    topographic, building site, 4
    underground survey, 18
    working, 4
Materials for color topography, 125
Mercator's map projection, 200
Meridian, 120

Method of indicating scales, 20
Military grid, 205
Military symbols, 69–76
Mountain formations, 154
Mounting display maps, 216
Mounting maps, on cloth, 217
    on paper, 217

N

Natural formations, picturing, 234
Nautical chart symbols, 57
North points, 119–120

O

Oblique sketching, 42
Office practice, 219
Oil-well symbols, 77
Orthographic map projection, **198**
Ozalid dry-point process, **187**

P

Pantagraph, use of, 180
Paper, qualities drawing paper should have, 230
    stretching for color work, 127
    types for drawing purposes, 232
    for water colors, 127
Parallel ruler, 209

Pencil transfer, 179
Penciling contours, 108
Penciling details, 107
Pencils, colored, use of, 139
Pens, choice and use of, 26
    rules for sharpening, 213
    special, 210
    use in contouring, 39
Perspective projection, reduction by, 184
Photostat, 187
Picturing natural formations, 234
Plan site, 157
Planimeter, polar, 223
    constant determination, 226
    theory of, 223
    use of, 227
    zero circle, 228
    radial, 230
    rolling, 230
Plotting, of contours from profiles, 160
    of points, 90
    of stadia notes, 103
    of traverses, 91
Points, location in field, 90
    position of, 90
Problems, in interpretation, 237
    from topographic sheets, 239
    visibility, 169, 170
Profile, from contour map, 236
    from plan, construction of, 157

Projections, map, 195
    azimuthal, 197
    conformal, 196
    conic, 201
    cylindrical, 199
    equal area, 196
    gnomonic, 198
    Lambert conformal, 203
    Mercator, 200
    orthographic, 198
    polyconic, 203
    states used in, 206
    transverse Mercator, 201
    types, 197
Proportional dividers, use of, 210
Proportional squares, 181
Protractors, circular, 103–105
    rectangular, 106
    use of two, 105

Q

Quadrangle sheets, U.S. Geological Survey, 154

R

Radial planimeter, 230
Rectangular coordinates, plotting traverse with, 98
Reduction of maps, 180–181, 184, 187

Relief, definition of, 2
Relief features, interpretation of, 237
Relief sketching, 108
Relief symbols, 38, 40, 44, 50–51
Reproduction, of city maps, 189
    mapping for, 192
Rock dragging, 132
Rolling planimeter, 230
Route location, 165
Ruler, parallel, 209
Rules, for colorwork, 127
    inking for reproduction, 194
    for scaling drawings, 209

S

Scale, bar, 22
    conversion, 21
    diagonal, 23
Scales, kinds of, 209
    map, 18
    methods of indicating, 20
    relation to contour interval, 19
    shading, 42–43
    table for general practice, 20
Sections, geological, 84–85
    standard A.R.E.A., 68
Sensitized tracing cloth, 186
Shading of slopes, 136
Side elevation from contour map. 163
Signs, conventional, discussion, 25

Site plan, 157
Sketching contours from control points, 160
Squares, proportional, 181–182
Stereographic map projection, 197
Stippling, 132
Straight edges, 208
Symbols, air navigation, 79
   for color topography, (*facing*), 139
   culture, 52–54, 61–67, 80
   Forest Service, 58
   geological, 81
   golf course, 78
   hydrographic, 46, 55–57
   military, 69–76
   nautical chart, 57
   oil and gas well, 77
   relief, 38, 40, 44, 50–51
   rules for making, 27
   shading of, 25
   showing character of soils, 35
   showing surface forms of ground, 39
   underground survey, 59
   vegetation, 48–49

T

Tinting a drawing, 216
Tints, blended, 131–132
   flat, 129
   graded, 131
   precautions in laying, 131
   preparation of, 128
   retouching, 130
   suggestions for laying, 137
Titles for topographic maps, 117
Topographic expression, limits of, 144
Topographic forms, glossary of, 172
Topographic information, classification, 1
Topographic maps, lettering for, 111
Topographic sheets, problems from, 239
Topographic sketching in oblique, 42
Tracing cloth, sensitized, 186
Tracing paper or cloth, 181
Tracings, rules for making, 183
Transfer by rubbing, 179
Transverse Mercator map projection, 201
Traverse plotting, by chords, 93
   by coordinates, 98
   latitudes and departures, 98

Traverse plotting, by polar coordinates, 91
   sines, 96
   tangents, 93
   tangents and parallel ruler, 96
Triangles, lettering, 220
Type letters, 193

U

Universal drafting machine, 221

V

Van Dyke paper, use of, 186
Vegetation, 2
Vegetation symbols, 48–49
Visibility problems, 169

W

Water colors, 125
Waterlining, 37

Z

Zinc etchings, 187